Validation in Chemical Measurement

Paul De Bièvre · Helmut Günzler (Eds.)

Validation in Chemical Measurement

Springer

Prof. Dr. Paul De Bièvre
Duineneind 9
2460 Kasterlee
Belgium

Prof. Dr. Helmut Günzler
Bismarckstrasse 11
69469 Weinheim
Germany

Library of Congress Control Number: 2004114858

ISBN 3-540-20788-0 **Springer Berlin Heidelberg New York**

Springer is a part of Springer Science+Business Media
springeronline.com

Cover design: *design & production* GmbH, Heidelberg
Production: LE-TEX Jelonek, Schmidt & Vöckler GbR, Leipzig

Printed on acid-free paper 52/3111/YL - 5 4 3 2 1 0

Preface

Validation of measurement methods has been used for a very long time in chemistry. It is mostly based on the examination of a measurement procedure for its characteristics such as precision, accuracy, selectivity, sensitivity, repeatability, reproducibility, detection limit, quantification limit and more.

When focussing on quality comparability and reliability in chemical measurement, the fields of interest to this Journal, one stumbles into various interpretations of the term *validation*. It is one more example of a term which is used sometimes very consistently, sometimes very loosely or indeed ambiguously. Since the term is very common in the chemical community, it is important that its meaning be clear. Turning to the 2nd edition of the International Vocabulary of Basic and General terms in Metrology (VIM) (1993), surprisingly we do not find a definition. Webster's Dictionary of the English language (1992) tells us that validation is 'making or being made valid'. Obviously *validation* has to do with *valid.* The same Webster indicates the meaning of the corresponding verb: to validate seems 'to make valid or binding, to confirm the validity of (Latin: validare)', where *valid* means: 'seen to be in agreement with the facts or to be logically sound'. We certainly can build on this to have a 'valid' discussion. Validation of a method clearly seems to mean making 'valid' the measurement results obtained by this method. The first definition 'seen to be in agreement with the facts', is rather difficult to apply. The second definition however, tells us that 'validation of a method is to make the method to be seen as logically sound'. It looks as if validation of a method is a process whereby it is tested and demonstrated by somebody or some authority to be logically sound. Such a validation should enable everybody to use it. That implies a list of methods 'validated' by competent authorities in the field concerned, which sounds possible and useful. Is that not what AOAC does?

Sometimes, the notion of validating a measurement result also shows up. Apparently it means to make a result 'valid', and even binding, i.e. confirming its 'validity'. Since *valid* means 'seen to be in agreement with the facts', that almost sounds as a synonym for 'accurate'. That makes sense and there seems to be no argument as to whether a method or a result can be validated (they can). An important question arises: does a validated method automatically give a validated measurement result, i.e. a quantity value[1] with associated measurement uncertainty? The answer must be: no. There can never be a mechanism or recipe for producing *automatically* 'valid' results because one can never eliminate the skills, the role and the responsibility of the analyst.

ISO 9000:2000, item 3.8.5 defines *validation* as 'confirmation by examination and provision of objective evidence that the requirements for an intended use are fulfilled'. The revised edition of the VIM ('VIM3'), is likely to fine-tune this definition of the concept 'validation' to be 'confirmation through examination of a given item and provision of objective evidence that it fulfills the requirements for a stated intended use'.

Looking at simple practice, many people are looking for a formal decision that a given measurement method *automatically* gives them '*valid*' i.e. reliable results. One wonders what this has to do with 'stated intended use'. Reliability clearly is a property of a measurement result. Checking whether that result fulfills the requirement for a stated intended use, seems to be a totally different matter. That requires the formulation of a requirement a priori, i.e. before the measurement is made, and derived from the need for a measurement result, not from the result itself.

This anthology contains 31 outstanding papers published in the Journal "Accreditation and Quality Assurance" since its inception, but mostly in the period 2000–2003, on the topic 'validation'. They reflect the latest understanding – or lack thereof –, of the concept and possibly some rationale(s) for the answer to the question why it is important to integrate the concept of 'validation' into the standard procedures of every measurement laboratory.

It is hoped that this anthology is of benefit to both the producers and the users of results of chemical measurements: the basic concepts and the basic thinking in measurement are the same for both.

Prof. Dr. P. De Bièvre
Editor-in-Chief
Accreditation and Quality Assurance
Kasterlee 2004-04-02

[1] quantity (German: 'Messgrösse', French: 'grandeur de mesure', Dutch: 'meetgrootheid') is not used here in the meaning 'amount', but as the generic term for the quantities we measure: concentration, volume, mass, temperature, time, etc., as defined in the VIM.

Contents

Bioanalytical method validation and its implications for forensic and clinical toxicology – A review 1

Validation of a computer program for atomic absorption analysis 10

Instrumental validation in capillary electrophoresis and checkpoints for method validation 14

Qualification and validation of software and computer systems in laboratories. Part 1: Validation during development 24

Clinical reference materials for the validation of the performance of photometric systems used for clinical analyses 31

Measurement uncertainty and its implications for collaborative study method validation and method performance parameters 37

Qualification and validation of software and computer systems in laboratories. Part 2: Qualification of vendors 42

Qualification and validation of software and computer systems in laboratories. Part 3: Installation and operational qualification 46

Qualification and validation of software and computer systems in laboratories. Part 4: Evaluation and validation of existing systems 51

Analytical validation in practice at a quality control laboratory in the Japanese pharmaceutical industry 56

Validation of the uncertainty evaluation for the determination of metals in solid samples by atomic spectrometry 62

Validation requirements for chemical methods in quantitative analysis – horses for courses? 68

Validation criteria for developing ion-selective membrane electrodes for analysis of pharmaceuticals .. 73

Is the estimation of measurement uncertainty a viable alternative to validation? 76

Validation of an HPLC method for the determination of p-dichlorobenzene and naphthalene in mothrepellents .. 80

The evaluation of measurement uncertainty from method validation studies. Part 1: Description of a laboratory protocol 84

The evaluation of measurement uncertainty from method validation studies. Part 2: The practical application of a laboratory protocol 91

Automated ion-selective measurement of lithium in serum. A practical approach to result-level verification in a two-way method validation 101

Determination of hydrocarbons in water – interlaboratory method validation before routine monitoring 107

Interlaboratory quality audit program for potable water – assessment of method validation done on inductively coupled plasma atomic emission spectrometer (ICP-AES) 112

A need for clearer terminology and guidance in the role of reference materials in method development and validation 116

Validation of salt spray corrosion test 121

Method validation and reference materials 128

Method validation of modern analytical techniques ... 134

Validation of test methods. General principles and concepts 139

Marketing Valid Analytical Measurement 143

Analytical procedure in terms of measurement (quality) assurance 148

Flexibilisation of the scope of accreditation: an important asset for food and water microbiology laboratories 153

Different approaches to legal requirements on validation of test methods for official control of foodstuffs and of veterinary drug residues and element contaminants in the EC 159

Methods validation: AOAC's three validation systems 163

Observing validation, uncertainty determination and traceability in developing Nordtest test methods 167

Contributors

Hassan Y. Aboul-Enein
Biological and Medical Research Department (MBC-03), King Faisal Specialist Hospital & Research Centre, P.O. Box 3354, Riyadh 11211, Saudi Arabia

G. Anand Kumar
VIMTA Labs Ltd., Hyderabad–500 051, India

Boris Anisimov
The National Physical Laboratory of Israel, Dancinger A. Bldg., Givat Ram, Jerusalem 91904, Israel

Elke Anklam
European Commission, Directorate General Joint Research Centre, Institute for Reference Materials and Measurements, Food Safety and Quality Unit, Retieseweg 111, B-2440 Geel, Belgium

Vicki J. Barwick
Laboratory of the Government Chemist, Queens Road, Teddington, Middlesex, TW11 0LY, UK

Ricardo J. N. Bettencourt da Silva
CECUL, Faculdade de Ciências da Universidade de Lisboa, P-1700 Lisbon, Portugal

M. Buzoianu
National Institute of Metrology, Sos. Vitan-Bârzesti 11, 75669 Bucharest, Romania

M. Filomena G. F. C. Camões
CECUL, Faculdade de Ciências da Universidade de Lisboa, P-1700 Lisbon, Portugal

J. J. Cuello
University of Zaragoza, Analytical Chemistry Department. Crtra. Zaragoza, s/n 22071 Huesca, Spain

R. C. Díaz
University of Zaragoza, Analytical Chemistry Department. Crtra. Zaragoza, s/n 22071 Huesca, Spain

Stephen L. R. Ellison
Laboratory of the Government Chemist, Queens Road, Teddington, Middlesex, TW11 0LY, UK

Rattanjit S. Gill
Laboratory of the Government Chemist, Queens Road, Teddington, Middlesex, TW11 0LY, UK

Alison Gillespie
Laboratory of the Government Chemist, Queens Road, Teddington, Middlesex, TW11 0LY, UK

Mykolas Gladkovas
Institute of Chemistry, Goštauto, 92600 Vilnius, Lithuania

H. Hey
Official food and veterinary laboratory of Schleswig-Holstein, 24537 Neumünster, Germany

D. Brynn Hibbert
Department of Analytical Chemistry, University of New South Wales, Sydney, NSW 2052, Australia

Magnus Holmgren
SP Swedish National Testing and Research Institute, P.O. Box 857, SE-50115 Borås, Sweden

Ludwig Huber
Hewlett-Packard GmbH Waldbronn, P.O. Box 1280, 76337 Waldbronn, Germany

Rouvim Kadis
D. I. Mendeleyev Institute for Metrology, 19 Moskovsky pr., 198005 St. Petersburg, Russia

H. Kohno
Quality Assurance Department, Sandoz Pharmaceuticals Ltd., No. 1 Takeno, Kawagoe City, Saitama, Japan

Reinhard Kreßner
Federal Environmental Agency, P. O. Box 33 00 22, 14191 Berlin, Germany

Stephan Küppers
Schering AG, In-Process-Control, Müllerstrasse 170–178, 13342 Berlin, Germany

Ilya Kuselman
The National Physical Laboratory of Israel, Dancinger A. Bldg., Givat Ram, Jerusalem 91904, Israel

Margreet Lauwaars
European Commission, Directorate General Joint Research Centre, Institute for Reference Materials and Measurements, Food Safety and Quality Unit, Retieseweg 111, B-2440 Geel, Belgium

Alexandre Leclercq
Institut Pasteur de Lille, Department of Food Microbiology and Safety, Rue du Professeur Calmette 1, 59000 Lille, France

Alex Lepek
Tech Projects, P.O. Box 9769, Jerusalem 91091, Israel

Peter Lepom
Federal Environmental Agency, P. O. Box 33 00 22, 14191 Berlin, Germany

Solveig Linko
Medix Diacor Labservice, Department of Quality Assurance, Nihtisillankuja 1, 00260 Espoo, Finland

Nineta Majcen
Metrology Institute of the Republic of Slovenia (MIRS), Laboratory Centre Celje, Tkalska 15, SI-3000 Celje, Slovenia

Hans H. Maurer
Department of Experimental and Clinical Toxicology, Institute of Experimental and Clinical Pharmacology and Toxicology, University of Saarland, 66421 Homburg, Germany

S. Othake
Quality Assurance Department, Sandoz Pharmaceuticals Ltd., No. 1 Takeno, Kawagoe City, Saitama, Japan

Frank T. Peters
Department of Experimental and Clinical Toxicology, Institute of Experimental and Clinical Pharmacology and Toxicology, University of Saarland, 66421 Homburg, Germany

Mark J. Q. Rafferty
Laboratory of the Government Chemist, Queens Road, Teddington, Middlesex, TW11 0LY, UK

Eugenija Ramoškienė
Institute of Chemistry, Goštauto, 92600 Vilnius, Lithuania

C. Ríos
University of Zaragoza, Analytical Chemistry Department. Crtra. Zaragoza, s/n 22071 Huesca, Spain

Gordon A. Ross
Hewlett-Packard GmbH Waldbronn, 76337 Waldbronn, Germany

H. S. Sahoo
VIMTA Labs Ltd., Hyderabad–501 051, India

Mudis Šalkauskas
Institute of Chemistry, Goštauto, 92600 Vilnius, Lithuania

M. A. Sarasa
University of Zaragoza, Analytical Chemistry Department. Crtra. Zaragoza, s/n 22071 Huesca, Spain

S. K. Satpathy
VIMTA Labs Ltd., Hyderabad–502 051, India

João Seabra e Barros
Instituto Nacional de Engenharia e Tecnologia Industrial, Estrada do Paço do Lumiar, P-1699 Lisbon Codex, Portugal

S. Seno
Quality Assurance Department, Sandoz Pharmaceuticals Ltd., No. 1 Takeno, Kawagoe City, Saitama, Japan

Avinoam Shenhar
The National Physical Laboratory of Israel, Dancinger A. Bldg., Givat Ram, Jerusalem 91904, Israel

Raluca-Iona Stefan
Department of Analytical Chemistry, Faculty of Chemistry, University of Bucharest, Blvd. Republicii No. 13, 70346 Bucharest-3, Romania

G. Swarnabala
VIMTA Labs Ltd., Hyderabad–503 051, India

Sue Upton
Laboratory of the Government Chemist, Queens Road, Teddington, Middlesex, TW11 0LY, UK

Robert J. Wells
Australien Government Analytical Laboratories, 1 Suakin Street, Pymble, NSW 2073, Australia

Herbert Wiederoder
Hewlett-Packard GmbH Waldbronn, P. O. Box 1280, 76337 Waldbronn, Germany

Alex Williams
19 Hamesmoor Way, Mytchett, Camberley, Surrey, GU16 6JG, UK

Peter Woitke
Federal Environmental Agency, P. O. Box 33 00 22, 14191 Berlin, Germany

Accred Qual Assur (2002) 7:441–449
DOI 10.1007/s00769-002-0516-5

Frank T. Peters
Hans H. Maurer

Bioanalytical method validation and its implications for forensic and clinical toxicology – A review

Received: 16 June 2002
Accepted: 12 July 2002

Part of this review was published in the communications of the International Association of Forensic Toxicologists (TIAFT; TIAFT Bulletin 32 (2002): 16–23) and of the Society for Forensic and Toxicologic Chemistry (GTFCH; Toxichem and Krimitech 68 (2001): 116-126).

F.T. Peters (✉) · H.H. Maurer
Department of Experimental and Clinical Toxicology,
Institute of Experimental and Clinical Pharmacology and Toxicology,
University of Saarland, 66421 Homburg, Germany
e-mail: frank.peters@uniklinik-saarland.de
Tel.: +49-6841-1626050
Fax: +49-6841-1626051

Abstract The reliability of analytical data is very important to forensic and clinical toxicologists for the correct interpretation of toxicological findings. This makes (bio)analytical method validation an integral part of quality management and accreditation in analytical toxicology. Therefore, consensus should be reached in this field on the kind and extent of validation experiments as well as on acceptance criteria for validation parameters. In this review, the most important papers published on this topic since 1991 have been reviewed. Terminology, theoretical and practical aspects as well as implications for forensic and clinical toxicology of the following validation parameters are discussed: selectivity (specificity), calibration model (linearity), accuracy, precision, limits, stability, recovery and ruggedness (robustness).

Keywords Method · Validation · Bioanalysis · Toxicology

Introduction

The reliability of analytical findings is a matter of great importance in forensic and clinical toxicology, as it is a prerequisite for correct interpretation of toxicological findings. Unreliable results might not only be contested in court, but could also lead to unjustified legal consequences for the defendant or to wrong treatment of the patient. The importance of validation, at least of routine analytical methods, can therefore hardly be overestimated. This is especially true in the context of quality management and accreditation, which have become matters of increasing importance in analytical toxicology in recent years. This is also reflected in the increasing requirements of peer-reviewed scientific journals concerning method validation. Therefore, this topic should be extensively discussed on an international level to reach a consensus on the extent of validation experiments and on acceptance criteria for validation parameters of bioanalytical methods in forensic and clinical toxicology.

Over the last decade, similar discussions have been going on in the closely related field of pharmacokinetic studies for registration of pharmaceuticals. This is reflected by the number of publications on this topic published in the last decade, of which the most important are discussed here.

Important publications on validation (1991 to present)

A review on validation of bioanalytical methods was published by Karnes et al. in 1991 which was intended to provide guidance for bioanalytical chemists [1]. One year later, Shah et al. published their report on the conference on "Analytical Methods Validation: Bioavailability, Bioequivalence and Pharmacokinetic Studies" held in Washington in 1990 (Conference Report) [2]. During this conference, consensus was reached on which parameters of bioanalytical methods should be evaluated, and some acceptance criteria were established. In the following years, this report was actually used as guidance by bioanalysts. Despite the fact, however, that some principle questions had been answered during this conference, no specific recommendations on practical issues like experimental designs or statistical evaluation were made. In 1994, Hartmann et al. analysed the Conference Report performing statistical experiments on the established acceptance criteria for accuracy and precision [3]. Based on their results they questioned the suitability of these criteria for practical application. From 1995 to 1997, application issues like experimental designs and statistical methods for bioanalytical method validation were discussed in a number of publications by Dadgar et al. [4, 5], Wieling et al. [6], Bressolle et al. [7] and Causon [8]. An excellent review on validation of bioanalytical chromatographic methods was published by Hartmann et al. in 1998, in which theoretical and practical issues were discussed in detail [9]. In an update of the Washington Conference in 2000, experiences and progress since the first conference were discussed. The results were again published by Shah et al. in a report (Conference Report II) [10], which has also been used as a template for guidelines drawn up by the U.S. Food and Drug Administration (FDA) for their own use [11]. Besides, it should be mentioned that some journals like the Journal of Chromatography B [12] or Clinical Chemistry have established their own criteria for validation. Two other documents that seem to be important in this context have been developed by the International Conference on Harmonisation of Technical Requirements for Registration of Pharmaceuticals for Human Use (ICH) and approved by the regulatory agencies of the European Union, the United States of America and Japan. The first, approved in 1994, concentrated on the theoretical background and definitions [13], the second, approved in 1996, on methodology and practical issues [14]. Both can be downloaded from the ICH homepage free of charge (www.ich.org). Finally, in 2001 Vander Heyden et al. published a paper on experimental designs and evaluation of robustness/ruggedness tests [15]. Despite the fact that the three last mentioned publications were not especially focussed on bioanalytical methods, they still contain helpful guidance on some principal questions and definitions in the field of analytical method validation.

The aim of our review is to present and compare the contents of the above mentioned publications on (bio)analytical method validation, and to discuss possible implications for forensic and clinical toxicology.

Terminology

The first problem encountered when studying literature on method validation are the different sets of terminology employed by different authors. A detailed discussion of this problem can be found in the review of Hartmann et al. [9]. Therein, it was proposed to adhere, in principle, to the terminology established by the ICH [13], except for accuracy, for which the use of a more detailed definition was recommended (cf. Accuracy). However, the ICH terminology lacked a definition for stability, which is an important parameter in bioanalytical method validation. Furthermore, the ICH definition of selectivity did not take into account interferences that might occur in bioanalysis (e.g. from metabolites). For both parameters, however, reasonable definitions were provided by Conference Report II [10].

Validation parameters

There is a general agreement that at least the following validation parameters should be evaluated for quantitative procedures: selectivity, calibration model (linearity), stability, accuracy (bias, precision) and limit of quantification. Additional parameters which might have to be evaluated include limit of detection, recovery, reproducibility and ruggedness (robustness) [2, 4–10, 12].

Selectivity (specificity)

In Conference Report II, selectivity was defined as follows: "Selectivity is the ability of the bioanalytical method to measure unequivocally and to differentiate the analyte(s) in the presence of components, which may be expected to be present". Typically, these might include metabolites, impurities, degradants, matrix components, etc. [10]. This definition is very similar to the one established by the ICH [13], but takes into account the possible presence of metabolites, and thus is more applicable for bioanalytical methods.

There are two points of view on when a method should be regarded as selective. One way to establish method selectivity is to prove the lack of response in blank matrix [1, 2, 4–10, 12, 14]. The requirement established by the Conference Report [2] to analyse at least six different sources of blank matrix has become state of the art. However, this approach has been subject to criticism in the review of Hartmann et al., who stated from statistical considerations, that relatively rare interferences will remain undetected with a rather high probability [9]. For the same reason, Dadgar et al. proposed to evaluate at least 10–20 sources of blank samples [4]. However, in Conference Report II [10], even analysis of only one source of blank matrix was deemed acceptable, if hyphenated mass spectrometric methods are used for detection.

The second approach is based on the assumption that small interferences can be accepted as long as precision and bias remain within certain acceptance limits. This approach was preferred by Dadgar et al. [4] and Hartmann et al. [9]. Both publications proposed analysis of up to 20 blank samples spiked with analyte at the lower limit of quantification (LLOQ) and, if possible, with interferents at their highest likely concentrations. In this approach, the method can be considered sufficiently selective if precision and accuracy data for these LLOQ samples are acceptable. For a detailed account of experimental designs and statistical methods to establish selectivity see Ref. [4].

Whereas the selectivity experiments for the first approach can be performed during a pre-validation phase (no need for quantification), those for the second approach are usually performed together with the precision and accuracy experiments during the main validation phase.

At this point it must be mentioned that the term specificity is used interchangeably with selectivity, although in a strict sense specificity refers to methods which produce a response for a single analyte, whereas selectivity refers to methods that produce responses for a number of chemical entities, which may or may not be distinguished [1]. Selective multi-analyte methods (e.g. for different drugs of abuse in blood) should of course be able to differentiate all interesting analytes from each other and from the matrix.

Calibration model (linearity)

The choice of an appropriate calibration model is necessary for reliable quantification. Therefore, the relationship between the concentration of analyte in the sample and the corresponding detector response must be investigated. This can be done by analysing spiked calibration samples and plotting the resulting responses versus the corresponding concentrations. The resulting standard curves can then be further evaluated by graphical or mathematical methods, the latter also allowing statistical evaluation of the response functions.

Whereas there is general agreement that calibration samples should be prepared in blank matrix and that their concentrations must cover the whole calibration range, recommendations on how many concentration levels should be studied with how many replicates per concentration level differ significantly [5–10, 12]. In Conference Report II, a sufficient number of standards to define adequately the relationship between concentration and response was demanded. Furthermore, it was stated that at least five to eight concentration levels should be studied for linear and maybe more for non-linear relationships [10]. However, no information was given on how many replicates should be analysed at each level. The guidelines established by the ICH and those of the Journal of Chromatography B also required at least five concentration levels, but again no specific requirements for the number of replicates at each level were given [12, 14]. Causon recommended six replicates at each of six concentration levels, whereas Wieling et al. used eight concentration levels in triplicate [6, 8]. Based on studies by Penninckx et al. [16], Hartmann et al. proposed in their review to rather use fewer concentration levels with a greater number of replicates (e.g. four evenly spread levels with nine replicates) [9]. This approach not only allows the reliable detection of outliers, but also a better evaluation of the behaviour of variance across the calibration range. The latter is important for choosing the right statistical model for the evaluation of the calibration curve. The often used ordinary least squares model for linear regression is only applicable for homoscedastic data sets (constant variance over the whole range), whereas in case of heteroscedasticity (significant difference between variances at lowest and highest concentration levels) the data should mathematically be transformed or a weighted least squares model should be applied [6–10]. Usually, linear models are preferable but, if necessary, the use of non-linear models is not only acceptable but even recommended. However, more concentration levels are needed for the evaluation of nonlinear models than for linear models [2, 9, 10].

After outliers have been purged from the data and a model has been evaluated visually and/or by, e.g. residual plots, the model fit should also be tested by appropriate statistical methods [2, 6, 9, 10, 14]. The fit of unweighted regression models (homoscedastic data) can be tested by the ANOVA lack-of-fit test [6, 9]. A detailed discussion of alternative statistical tests for both unweighted and weighted calibration models can be found in Ref. [16]. The widespread practice to evaluate a calibration model via its coefficients of correlation or determination is not acceptable from a statistical point of view [9].

However, one important point should be kept in mind when statistically testing the model fit: The higher the precision of a method, the higher the probability to detect a statistically significant deviation from the assumed calibration model [1, 6, 9]. Therefore, the practical relevance of the deviation from the assumed model should also be taken into account. If the accuracy data (bias and precision) are within the required acceptance limits or an alternative calibration model is not applicable, slight deviations from the assumed model may be neglected [6, 9].

Once a calibration model has been established, the calibration curves for other validation experiments (precision, bias, stability, etc.) and for routine analysis can be prepared with fewer concentration levels and fewer or no replicates [6, 9].

Accuracy

The accuracy of a method is affected by systematic (bias) as well as random (precision) error components [3, 9] This fact has been taken into account in the definition of accuracy as established by the International Organization for Standardization (ISO) [17]. However, it must be mentioned that accuracy is often used to describe only the systematic error component, i.e. in the sense of bias [1, 2, 6–8, 10, 12, 13]. In the following, the term accuracy will be used in the sense of bias, which will be indicated in brackets.

Bias

According to ISO, bias is the difference between the expectation of the test results and an accepted reference value [17]. It may consist of more than one systematic error component. Bias can be measured as a percent deviation from the accepted reference value. The term trueness expresses the deviation of the mean value of a large series of measurements from the accepted reference value. It can be expressed in terms of bias.

Due to the high workload of analysing such large series, trueness is usually not determined during method validation, but rather from the results of a great number of quality control (QC) samples during routine application or in interlaboratory studies.

Precision

According to ICH, precision is the closeness of agreement (degree of scatter) between a series of measurements obtained from multiple sampling of the same homogenous sample under the prescribed conditions and may be considered at three levels: repeatability, intermediate precision and reproducibility [13]. Precision is usually measured in terms of imprecision expressed as an absolute or relative standard deviation and does not relate to reference values.

Repeatability. Repeatability expresses the precision under the same operating conditions over a short interval of time. Repeatability is also termed intra-assay precision [13]. Within-run or within-day precision are also often used to describe repeatability.

Intermediate precision. Intermediate precision expresses within-laboratories variations: different days, different analysts, different equipment, etc. [13]. The ISO definition used the term "M-factor different intermediate precision", where the M-factor expresses how many and which factors (time, calibration, operator, equipment or combinations of those) differ between successive determinations [17]. In a strict sense, intermediate precision is the total precision under varied conditions, whereas so called inter-assay, between-run or between-day precision only measure the precision components caused by the respective factors. However, the latter terms are not clearly defined and obviously often used interchangeably with each other and also with the term intermediate precision.

Reproducibility. Reproducibility expresses the precision between laboratories (collaborative studies, usually applied to standardization of methodology) [13]. Reproducibility only has to be studied, if a method is supposed to be used in different laboratories.

Unfortunately, some authors also used the term reproducibility for within-laboratory studies at the level of intermediate precision [8, 12]. This should, however, be avoided in order to prevent confusion.

As already mentioned above, precision and bias can be estimated from the analysis of QC samples under specified conditions. As both precision and bias can vary substantially over the calibration range, it is necessary to evaluate these parameters at least at three concentration levels (low, medium and high relative to the calibration range) [1, 2, 9, 10, 14]. In Conference Report II, it was further defined that the concentration of the low QC sample must be within three times LLOQ [10]. The Journal of Chromatography B requirement is to study precision and bias at two concentration levels (low and high), whereas in the experimental design proposed by Wieling et al. four concentration levels (LLOQ, low, medium, high) were studied [6, 12]. Causon also suggested to estimate precision at four concentration levels [8]. Several authors have specified acceptance limits for precision and/or accuracy (bias) [2, 7, 8, 10, 12]. Both Conference Reports required precision to be within 15% relative standard deviation (RSD) except at the LLOQ where 20% RSD was accepted. Bias was required to be within ±15% of the accepted true value, except at the LLOQ where ±20% were accepted [2, 10]. These requirements have been subject to criticism in the analysis of the Conference Report by Hartmann et al. [3]. They concluded from statistical considerations that it is not realistic to apply the same acceptance criteria at different levels of precision (repeatability, reproducibility) as RSD under reproducibility conditions is usually considerably greater than under repeatability conditions. Furthermore, if precision and bias estimates are close to the acceptance limits, the probability to reject an actually acceptable method (β-error) is quite high. Causon proposed the same acceptance limits of 15% RSD for precision and ±15% for accuracy (bias) for all concentration levels [8].

The guidelines established by the Journal of Chromatography B required precision to be within 10% RSD for the high QC samples and within 20% RSD for the low QC sample. Acceptance criteria for accuracy (bias) were not specified therein [12].

Again, the proposals on how many replicates at each concentration levels should be analysed vary considerably. The Conference Reports and Journal of Chromatography B guidelines required at least five replicates at each concentration level [2, 10, 12]. However, one would assume that these requirements only apply to repeatability studies; at least no specific recommendations were given for studies of intermediate precision or reproducibility. Some more practical approaches to this problem have been described by Wieling et al. [6], Causon [8] and Hartmann et al. [9]. In their experimental design, Wieling et al. analysed three replicates at each of four concentration levels on each of 5 days. Similar approaches were suggested by Causon (six replicates at each of four concentrations on each of four occasions) and Hartmann et al. (two replicates at each concentration level on each of 8 days). All three used or proposed one-way ANOVA to estimate repeatability and time-different precision components. In the design proposed by Hartmann et al. the degrees of freedom for both estimations are most balanced, namely 8 for within-run precision and 7 for between-run precision. In the information for authors of the Clinical Chemistry journal, an experimental design with two replicates per run, two runs per day over 20 days for each concentration level is recommended, which has been established by the NCCLS [18]. This not only allows estimation of within-run and between-run standard deviations, but also of within-day, between-day and total standard deviations, which are in fact all estimations of precision at different levels. However, it seems questionable if the additional information provided by this approach can justify the high workload and costs compared to the other experimental designs.

Daily variations of the calibration curve can influence bias estimation. Therefore, bias estimation should be based on data calculated from several calibration curves [9]. In the experimental design of Wieling et al., the results for QC samples were calculated via daily calibration curves. Therefore, the overall means from these results at the different concentration levels reliably reflect the average bias of the method at the corresponding concentration level. Alternatively, as described in the same paper, the bias can be estimated using confidence limits around the calculated mean values at each concentration [6]. If the calculated confidence interval includes the accepted true value, one can assume the method to be free of bias at a given level of statistical significance. Another way to test the significance of the calculated bias is to perform a *t*-test against the accepted true value.

However, even methods exhibiting a statistically significant bias can still be acceptable, if the calculated bias lies within previously established acceptance limits. Other methods for bias evaluation can be found in Ref. [9].

Limits

Lower limit of quantification (LLOQ)

The LLOQ is the lowest amount of an analyte in a sample that can be quantitatively determined with suitable precision and accuracy (bias) [10, 13]. There are different approaches for the determination of LLOQ.

LLOQ based on precision and accuracy (bias) data [2, 7–10, 13, 14]. This is probably the most practical approach and defines the LLOQ as the lowest concentration of a sample that can still be quantified with acceptable precision and accuracy (bias). In the Conference Reports, the acceptance criteria for these two parameters at LLOQ are 20% RSD for precision and ±20% for bias. Only Causon suggested 15% RSD and ±15%, respectively [8]. It should be pointed out, however, that these parameters must be determined using an LLOQ sample independent from the calibration curve. The advantage of this approach is the fact that the estimation of LLOQ is based on the same quantification procedure used for real samples.

LLOQ based on signal to noise ratio (S/N) [12, 14]. This approach can only be applied if there is baseline noise, e.g. to chromatographic methods. Signal and noise can then be defined as the height of the analyte peak (signal) and the amplitude between the highest and lowest point of the baseline (noise) in a certain area around the analyte peak. For LLOQ, S/N is usually required to be equal to or greater than 10.

The estimation of baseline noise can be quite difficult for bioanalytical methods, if matrix peaks elute close to the analyte peak.

LLOQ based on standard deviation of the response from blank samples [14]. Another definition of LLOQ is the concentration that corresponds to a detector response that is k times greater than the estimated standard deviation of blank samples (SD_{bl}). From the detector signal, the LLOQ can be calculated using the slope of the calibration curve (S) with following formula: $LLOQ = k \cdot SD_{bl}/S$ (for blank corrected signals).

This approach is only applicable for methods where SD_{bl} can be estimated from replicate analysis of blank samples. It is therefore not applicable for most quantitative chromatographic methods, as here the response is usually measured in terms of peak area units, which can of course not be measured in a blank sample analysed with a selective method.

LLOQ based on a specific calibration curve in the range of LLOQ [14]. In this approach, a specific calibration curve is established from samples containing the analyte in the range of LLOQ. One must not use the calibration curve over the whole range of quantification for this determination. The standard deviation of the blank can then be estimated from the residual standard deviation of the regression line or the standard deviation of the y intercept. The calculations of LLOQ are basically the same as described under the heading "LLOQ based on standard deviation of the response from the blank samples". This approach is also applicable for chromatographic methods.

Upper limit of quantification (ULOQ)

The upper limit of quantification is the maximum analyte concentration of a sample that can be quantified with acceptable precision and accuracy (bias). In general the ULOQ is identical to the concentration of the highest calibration standard [10].

Limit of detection (LOD)

Quantification below LLOQ is by definition not acceptable [2, 5, 9, 10, 13, 14]. Therefore, below this value a method can only produce semiquantitative or qualitative data. However, it can still be important to know the LOD of the method. According to ICH, it is the lowest concentration of analyte in a sample which can be detected but not necessarily quantified as an exact value. According to Conference Report II, it is the lowest concentration of an analyte in a sample, that the bioanalytical procedure can reliably differentiate from background noise [10, 13].

The approaches for estimation of the LOD are basically the same as those described for LLOQ under the headings "LLOQ based on signal to noise ratio (S/N)" – "LOQ based on a specific calibration curve in the range of LLOQ". However, for LOD a S/N or k-factor equal to or greater than three is usually chosen [1, 6, 9, 14]. If the calibration curve approach is used for determination of the LOD, only calibrators containing the analyte in the range of LOD must be used.

Stability

The definition according to Conference Report II was as follows: "The chemical stability of an analyte in a given matrix under specific conditions for given time intervals" [10]. Stability of the analyte during the whole analytical procedure is a prerequisite for reliable quantification. Therefore, full validation of a method must include stability experiments for the various stages of analysis including storage prior to analysis.

Long-term stability

The stability in the sample matrix should be established under storage conditions, i.e. in the same vessels, at the same temperature and over a period at least as long as the one expected for authentic samples [1, 2, 4, 5, 9, 10, 12].

Freeze/thaw stability

As samples are often frozen and thawed, e.g. for re-analysis, the stability of analyte during several freeze/thaw cycles should also be evaluated. The Conference Reports require a minimum of three cycles at two concentrations in triplicate, which has also been accepted by other authors [2, 4, 6, 9, 10].

In-process stability

The stability of analyte under the conditions of sample preparation (e.g. ambient temperature over time needed for sample preparation) is evaluated here. There is general agreement, that this type of stability should be evaluated to find out, if preservatives have to be added to prevent degradation of analyte during sample preparation [4, 9, 10].

Processed sample stability

Instability cannot only occur in the sample matrix, but also in prepared samples. It is therefore important to also test the stability of an analyte in the prepared samples under conditions of analysis (e.g. autosampler conditions for the expected maximum time of an analytical run). One should also test the stability in prepared samples under storage conditions, e.g. refrigerator, in case prepared samples have to be stored prior to analysis [4–6, 9, 10].

For more details on experimental design and statistical evaluation of stability experiments see Refs. [4, 5, 9].

Stability can be tested by comparing the results of QC samples analysed before (comparison samples) and after (stability samples) being exposed to the conditions for stability assessment. It has been recommended to perform stability experiments at least at two concentration levels (low and high) [4–6, 9]. For both, comparison and stability samples, analysis of at least six replicates was recommended [9]. Ratios between comparison samples and stability samples of 90–110% with 90% confidence intervals within 80–120% [9] or 85–115% [4] were regarded acceptable. Alternatively, the mean of the reference samples can be tested against a lower acceptance limit corresponding to 90% of the mean of the comparison samples [8, 9].

Recovery

As already mentioned above, recovery is not among the validation parameters regarded as essential by the Conference Reports. Most authors agree, that the value for recovery is not important, as long as the data for LLOQ, (LOD), precision and accuracy (bias) are acceptable [1, 5, 7–10]. It can be calculated as the percentage of the analyte response after sample workup compared to that of a solution containing the analyte at the theoretical maximum concentration. Therefore, absolute recoveries can usually not be determined if the sample workup includes a derivatization step, as the derivatives are often not available as reference substances. Nevertheless, the guidelines of the Journal of Chromatography B require the determination of the recovery for analyte and internal standard at high and low concentrations [12].

Ruggedness (robustness)

Ruggedness is a measure for the susceptibility of a method to small changes that might occur during routine analysis like small changes of pH values, mobile phase composition, temperature, etc. Full validation must not necessarily include ruggedness testing; it can however be very helpful during the method development/pre-validation phase, as problems that may occur during validation are often detected in advance. Ruggedness should be tested, if a method is supposed to be transferred to another laboratory [1, 9, 13–15]. A detailed account and helpful guidance on experimental designs and evaluation of ruggedness/robustness tests can be found in Ref. [15].

Implications for forensic and clinical toxicology

Almost all of the above mentioned publications referred to bioanalytical methods for bioavailability, bioequivalence or pharmacokinetic studies. This field is of course very closely related to forensic and clinical toxicology, especially if only routine methods are considered. Therefore, it seems reasonable to base the discussion concerning method validation in toxicological analysis on the experiences and consensus described above and not to start the whole discussion anew. In the following, possible implications for forensic and clinical toxicology will be discussed.

Terminology

As already mentioned above, there are several sets of terminology in the literature. It is therefore strongly recommended to adopt, in principle, one of these sets for validation in forensic and clinical toxicology and add slight modifications, where it seems necessary. The definitions established by the ICH seem to be a reasonable choice as they are consensus definitions of an international conference and easily available on the homepage of ICH (www.ich.org).

Validation parameters

Selectivity (specificity)

During pharmacokinetic studies (therapeutic) drugs are usually ingested under controlled conditions. Therefore, there is no need to prove the ingestion of this drug. Due to this fact the selectivity evaluation can be based on the acceptability of precision and accuracy data at the LLOQ. This approach is quite problematic for forensic and clinical toxicology, where analysis is often mainly performed to prove ingestion of an (illicit) substance and, therefore, qualitative data are also important. Here, the approach to prove selectivity by absence of signals in blank samples makes much more sense. The confinement of Conference Report II [10] to only study one source of blank matrix for methods employing MS detection does not seem reasonable for toxicological applications because of the great importance of selectivity in this field. However, discussion is needed on how many sources of blank samples should be analysed and if this should depend on the detection method.

It seems also reasonable to also check for interferences from other xenobiotics that can be expected to be present in authentic samples (e.g. other drugs of abuse for methods to determine MDMA, other neuroleptics for methods to determine olanzapine). This can be accomplished by spiking these possible interferents at their highest expectable concentrations into blank matrix and checking for interferences after analysis. Another way to exclude interferences from such compounds is to check authentic samples containing these but not the analyte for interfering peaks. This latter approach is preferable, if the possibly interfering substance is known to be extensively metabolized, as it also allows exclusion of interferences from such metabolites, which are usually not available as pure substances.

Calibration model

The use of matrix-based calibration standards seems also important in toxicological analysis, in order to account for matrix effects during sample workup and measurement (e.g. by chromatographic methods). Consensus should be reached on how many concentration levels and how many replicates per level should be analysed. From our own experience six levels with six replicates each seems reasonable. Weighted calibration models will generally be the most appropriate in toxicological analysis, as concentration ranges of analytes in toxicological samples are usually much greater than in samples for pharmacokinetic studies. Homoscedasticity, a prerequisite for unweighted models, can however only be expected for small calibration ranges.

Accuracy (precision and bias)

There is no obvious reason to evaluate these parameters differently than has been described above. Due to the often higher concentration ranges, it might be reasonable to also validate the analysis of QC samples containing concentrations above the highest calibration standard after dilution or after reduction of sample volumes, as it has been described by Wieling et al. [6] and Dadgar et al. [5]. The latter has also described the use of QC samples with concentrations below those of the lowest calibration standard using greater sample volumes.

Limits

The same approaches and criteria as those described above under "Limits" could be used. All approaches have been described to a lesser or greater extent in international publications, especially for the determination of LOD. Nevertheless, it seems important to reach consensus on this matter at least for forensic and clinical toxicology, as reliable detection of a substance is one of the most important issues in toxicological analysis. At this point it must be stressed that for the estimation of LOD and LLOQ via a special calibration curve, the calibration samples must only contain the analyte at concentrations close to LOD and LLOQ. Use of the calibration curve over the whole range may lead to overestimation of these limits.

Stability

The biggest problems encountered during stability testing for bioanalytical methods in forensic and clinical toxicology is the fact that there is a great number of different sampling vessels. Furthermore, the anticoagulants used also differ. Both facts make it difficult to assess long-term stability, as the workload to analyse all possible combinations of vessels and anticoagulants is of course far to great. However, for some analytes relevant to forensic and clinical toxicology (e.g. cocaine) stability problems with different sampling vessels have been reported [19]. Therefore, the relevance of this parameter to forensic and clinical toxicology has to be discussed extensively. Agreement on a single type of vessel to use for sampling of toxicological samples would probably be the easiest solution. Another problem is the fact that storage conditions prior to arrival in the laboratory are not known. So this matter will also have to be addressed.

Recovery

Recovery does not seem to be a big issue for forensic and clinical toxicologists as long as precision, accuracy (bias), LLOQ and, especially, LOD are satisfactory. However, during method development one should of course try to optimize recovery.

Ruggedness

There is no obvious reason to treat this matter differently than described above under "Ruggedness (robustness)".

Measurement uncertainty

Measurement uncertainty, a parameter characterizing the dispersion of the values that could reasonably be attributed to the measurand (e.g. concentration), is considered an important concept in analytical chemistry [20]. It comprises many components and is generally reported as a standard deviation or as a confidence interval. However, measurement uncertainty was not explicitly addressed in any of the publications on bioanalytical method validation discussed in this review. One probable reason might be that measurement uncertainties of modern analytical methods are certainly small compared to the differences encountered between individual subjects in pharmacokinetic studies. Nevertheless, knowledge about the uncertainty of measurements can be very helpful or even important for the interpretation of bioanalytical data, especially in forensic toxicology. As bioanalytical methods are usually rather complex and assessment of the contribution of individual components on the combined uncertainty of the results would therefore be time consuming and costly, estimation of measurement uncertainty from validation data, especially precision and accuracy data, would be preferable. The more the design of validation experiments accounts for conditions during routine application of a method, i.e. the more individual components are comprised in one validation parameter, the better the estimation of measurement uncertainty from those parameters. Consequently, reproducibility data from interlaboratory experiments studies or intermediate precision data with M-factors ≥ 3 from intralaboratory experiments should adequately reflect measurement uncertainty of bioanalytical methods. However, further components not covered adequately by validation experiments might play a role for measurement uncertainty. A detailed guide on the quantification of analytical measurement uncertainty has been published by EURACHEM/CITAC [20].

Conclusion

There are only a few principle differences concerning validation of bioanalytical methods in the fields of pharmacokinetic studies and forensic and clinical toxicology. Therefore, it seems reasonable to base the discussion on validation in the field of toxicology on the experiences and consensus already existing in the closely related field of pharmacokinetic studies for registration of pharmaceuticals and focus the discussion on those parameters, which are of special importance for toxicologists, i.e. selectivity, LOD, LLOQ and stability.

References

1. Karnes HT, Shiu G, Shah VP (1991) Pharm Res 8:421–426
2. Shah VP, Midha KK, Dighe S, McGilveray IJ, Skelly JP, Yacobi A, Layloff T, Viswanathan CT, Cook CE, McDowall RD, Pittman KA, Spector S (1992) Pharm Res 9:588–592
3. Hartmann C, Massart DL, McDowall RD (1994) J Pharm Biomed Anal 12:1337–1343
4. Dadgar D, Burnett PE (1995) J Pharm Biomed Anal 14:23–31
5. Dadgar D, Burnett PE, Choc MG, Gallicano K, Hooper JW (1995) J Pharm Biomed Anal 13:89–97
6. Wieling J, Hendriks G, Tamminga WJ, Hempenius J, Mensink CK, Oosterhuis B, Jonkman JH (1996) J Chromatogr A 730:381–394
7. Bressolle F, Bromet PM, Audran M (1996) J Chromatogr B 686:3–10
8. Causon R (1997) J Chromatogr B 689:175–180
9. Hartmann C, Smeyers-Verbeke J, Massart DL, McDowall RD (1998) J Pharm Biomed Anal 17:193–218
10. Shah VP, Midha KK, Findlay JW, Hill HM, Hulse JD, McGilveray IJ, McKay G, Miller KJ, Patnaik RN, Powell ML, Tonelli A, Viswanathan CT, Yacobi A (2000) Pharm Res 17:1551–1557
11. U.S. Department of Health and Human Services, Food and Drug Administration (2001) Guidance for industry, bioanalytical method validation http://www.fda.gov/cder/guidance/index.htm
12. Lindner W, Wainer IW (1998) J Chromatogr B 707:1–2
13. International Conference on Harmonization (ICH) (1994) Validation of analytical methods: Definitions and terminology ICH Q2 A
14. International Conference on Harmonization (ICH) (1996) Validation of analytical methods: Methodology ICH Q2 B
15. Vander-Heyden Y, Nijhuis A, Smeyers-Verbeke J, Vandeginste BG, Massart DL (2001) J Pharm Biomed Anal 24:723–753
16. Penninckx W, Hartmann C, Massart DL, Smeyers-Verbeke J (1996) J Anal At Spectrom 11:237–246
17. International Organization for Standardization (ISO) (1994) Accuracy (trueness and precision) of measurement methods and results: ISO/DIS 5725-1 to 5725-3. ISO, Geneva, Switzerland
18. NCCLS (1999) Evaluation of precision performance of clinical chemistry devices; Approved guideline. NCCLS, Wayne, Pa., USA
19. Toennes SW, Kauert GF (2001) J Anal Toxicol 25:339–343
20. EURACHEM/CITAC (2000) Quantifying uncertainty in analytical measurement http://www.eurachem.bam.de/guides/quam2.pdf

Accred Qual Assur (1997) 2:234–237

Ilya Kuselman
Boris Anisimov
Avinoam Shenhar
Alex Lepek

Validation of a computer program for atomic absorption analysis

Received: 25 January 1997
Accepted: 14 March 1997

I. Kuselman (✉) · B. Anisimov
A. Shenhar
The National Physical laboratory of Israel, Danciger A. Bldg, Givat Ram, Jerusalem 91904, Israel
Tel: +972-2-6536534
Fax: +972-2-6520797
e-mail: freddy@vms.huji.ac.il

A. Lepek
Tech Projects, P. O. Box 9769, Jerusalem 91091, Israel

Abstract The approach to validation of a computer program for an analytical instrument as a component of the analytical method (using this instrument with the program) is discussed. This approach was used for validating a new program for atomic absorption analysis. The validation plan derived from this approach was based on minimising the influence of all steps of the analytical procedure on the analytical results obtained by the method. In this way significant changes in the results may be caused only by replacement of the previous program by the new one. The positive validation conclusion was based on the comparison of the results of the analysis of suitable reference materials obtained with the new program and with its precursor in the same conditions, and also on comparison of their deviations from the accepted reference values for these materials, with the corresponding uncertainties.

Key words Validation · Computer program · Analytical instrument · Measurement uncertainty · Atomic absorption analysis

Introduction

The validation of analytical methods is a well-known problem in the analytical community [1]. The international guidance for equipment qualification (EQ) of analytical instruments and their validation is in the development stage [2–4]. At this time validation of computer systems for analytical instruments is less elaborated [5–7]. The term "computer system" comprises computer hardware, peripherals, and software that includes application programs and operating environments (MS-DOS, MS-Windows and others) [5, 6]. Since programs, software and the whole computer system are elements of the instrument used by the analyst according to the analytical method, successful validation of the method as a "black box" [8] means successful validation of the instrument, computer system, software and programs. On the other hand, the same instrument may also be calibrated and validated as a smaller (included) "black box" [9], with a corresponding validation conclusion about the computer system and its elements. Next, the computer system is again a smaller included "black box" [10, 11] etc. It is like a matreshka – a Russian wooden doll with successively smaller ones fitted into it. Therefore the approach to the validation of a computer program as a validation of a component of the analytical method is sound. In the framework of this approach, not all the method validation parameters (performance characteristics [10, 11]) may be relevant for the program validation. For example, the validation parameter "accuracy" (bias) is meaningful in this context, while the "selectivity", "specificity" or "ruggedness" are not. The bias may be evaluated as the difference between the results of the analysis of suitable reference materials obtained with the new program and with its precursor in the same conditions, and also by evaluation of their deviation from the accepted reference values for these materials in comparison to the corresponding uncertainties [12]. Certainly, the program validated in this way for the specific analytical method cannot be considered as validated for other applications (methods).

In this paper, the present approach is used for validating the new program GIRAF for atomic absorption analysis with the Perkin-Elmer 5000 spectrophotometer equipped with the graphite furnace and/or flame facilities. The program was developed by Tech Projects according to specifications of the National Physical Laboratory of Israel. The purpose of the development was to replace the out-of-date Perkin-Elmer HGA Graphics Software Package with Data System-10 (the operating environment is PETOS) and the obsolete Perkin-Elmer computer 3600 Data Station used with the Graphite furnace from 1983. Moreover, the technique for atomic absorption analysis with flame facilities included visual reading of absorbance or emission values from the spectrophotometer display that did not correspond to the Good Laboratory Practice requirements. In addition to Perkin-Elmer equipment, an IBM compatible PC with the program Quattro-Pro for Windows was used routinely by us for the linear regression analysis and uncertainty calculations. The same calculations were done with the raw data of the graphite furnace analysis (absorbance peak area or peak height) obtained from Perkin-Elmer HGA Graphics Software Package with Data System-10.

GIRAF program description

The operating computer environment is MS-DOS, and the hardware is an IBM-compatible PC. GIRAF reads, displays and stores data from the spectrophotometer (absorbance or emission) as a function of time at a rate of 50 readings per second. The raw data can be displayed after filtering. The filter is an exponential one with a time constant of 0.2 s. The stored data are used to calculate the calibration curve and the analysis results including uncertainties in accordance with [12]. The calculations are based on the linear regression of data with replicate measurements. If the graphite furnace is used, the absorbance peak area and peak height are calculated as the data for regression analysis; if the flame facilities are used, the average absorbance or emission over the defined time are the data. The program also provides for analysis using the standard addition method.

Validation plan

In spite of the replacement of all elements of the computer system, it is obvious that in our case only the programs can influence the analytical results. So, the plan of GIRAF validation consisted in comparison of the results of the analysis of the reference materials obtained with the program to be validated with the previous routinely used ones. The objects for analysis, commonly acceptable as the simplest, were chosen for minimising the influence of all the steps of the analytical procedure on the analytical results. These objects are aqueous solutions of lead for the graphite furnace, of copper for flame atomic absorption, and of sodium for flame atomic emission analysis. The preparation of the test solutions and the solutions for the spectrophotometer calibration (five in each calibration range) was planned to be done from the same sources. Differences between the results of the analysis obtained for the test solutions with the GIRAF program C_{new} and with its precursors (no change of any other conditions) C_{pre} and also their deviations from the accepted reference values for these solutions were planned as the data for the validation conclusion. The conclusion can be positive if these data correspond to the following acceptance criteria:

A. $|C_{new}-C_{pre}| \leq (U_{new}^2+U_{pre}^2)^{1/2}$, where U_{new} and U_{pre} are are the uncertainties (95% confidence) of C_{new} and C_{pre}, correspondingly;

B. $|C-C_{tr}| \leq (U^2+U_{tr}^2)^{1/2}$, where C and U are C_{new} and U_{new} or C_{pre} and U_{pre}, while C_{tr} and U_{tr} are the "true" concentration of the analyte in the reference material and its uncertainty (95% confidence).

Experimental

Reagents

Merck standard solutions were used: lead(II) nitrate in 0.5 M nitric acid, certified lead concentration 998±2 mg/L, shelf life up to 31. 03. 98; copper(II) nitrate in 0.5 M nitric acid, certified copper concentration 1000±2 mg/L, shelf life up to 31. 10. 98; sodium nitrate in water, certified concentration 1000±2 mg/L, shelf life up to 01. 06. 97. Nitric acid (1 vol%) as the solution for dilution of the standard solutions was prepared from Merck nitric acid Suprapur (65%) and deionised water.

$Mg(NO_3)_2 \cdot 6\,H_2O$ Suprapur, and $(NH_4)_2HPO_4$ Pro analysis, both from Merck, were used for matrix modifier preparation for lead determination by graphite furnace analysis.

CsCl Suprapur from Merck was used for matrix modifier preparation for sodium determination by flame atomic emission analysis.

Reference materials

The reference materials for calibration of the spectrophotometer with the graphite furnace system were prepared from the lead standard solution by its dilution with nitric acid 1% solution: blank (nitric acid 1% solution), 5, 10, 20, 30 and 40 μg/L (ppb). A solution with a lead concentration of 25 ppb ("true" value) was used as a test solution, i.e. a sample with "unknown" analyte concentration.

The reference materials for calibration of the spectrophotometer for flame atomic absorption analysis were prepared from the copper standard solution by its dilution with nitric acid 1% solution: blank (nitric acid 1% solution), 1, 2, 3, 4 and 5 mg/L (ppm). A solution with a copper concentration of 2.5 ppm ("true" value) was used as a test solution.

The reference materials for the calibration of the spectrophotometer for flame atomic emission analysis were prepared from the sodium standard solution by its dilution with water and addition of the matrix modifier: blank (CsCl 0.1% solution in water), 0.1, 0.2, 0.4, 0.6 and 1.0 ppm. A solution with sodium concentration of 0.5 ppm ("true" value) was used as a test solution.

It may be shown that, according to the preparation procedure of the test solutions, the relative expanded uncertainty [12] of the corresponding analyte concentrations is approximately 1% (rel.) at 95% confidence, i.e. the correct concentrations C_{tr} and their uncertainies U_{tr} are 25.0±0.3 ppb for lead, 2.50±0.03 ppm for copper, and 0.50±0.01 ppm for sodium.

Technique of the experiment

All measurements were performed with the program GIRAF and with the previous ones without shutdown of the instrument at the same analytical conditions (according to Perkin-Elmer "cookbook") by the same analyst with the same reference materials applied as calibration and test solutions. The matrix modifier for lead determination by graphite furnace analysis was used separately from the reference materials, i.e. was not mixed with them. The matrix modifier for sodium determination by flame atomic emission analysis was introduced in all corresponding solutions (the blank and the reference materials for the calibration and test). Three replicates of all measurements were made.

Results and discussion

Results of the analysis (C) are presented in Table 1. Usually the display of the Perkin-Elmer 5000 instrument in flame atomic absorption analysis is autozeroed for the blank solution. For this reason the values read visually from the display and those stored by GIRAF differ by the value of the blank, a fact that has no influence on the results of the analysis. The uncertainties of the results (U) are shown in Table 1 also. The uncertainties of the calculated analyte concentrations in the test solutions U_{tr} described in the Reference materials section are approximately one third of the corresponding uncertainties of the results of the analysis (for sodium it is not obvious because of rounding), and so the use of the term "true" values in the context of the validation is permissible. All these data were used for calculation of the acceptance criteria A and B formulated in the Validation plan (see Table 2). The criteria are satisfied for both graphite furnace and flame analysis.

It is worth while to note also that the program GIRAF shortens the analysis time and is more easy to use than the previous programs, but evaluation of these parameters was not included in the validation.

Table 1 Results of the experiment (average values of three replicates) for validation of the new program GIRAF. Conc., concentrations; Prev., previous; prog., program; Test, test solution

Graphite furnace					Flame					
Pb conc., ppb	Peak height Prev. prog.	Peak height New prog.	Peak area Prev. prog.	Peak area New prog.	Cu conc., ppm	Absorbance Prev. prog.	Absorbance New prog.	Na conc., ppm	Emission Prev. prog.	Emission New prog.
Blank	0.011	0.015	0.006	0.008	Blank	0.000	0.997	Blank	−0.004	0.103
5	0.056	0.060	0.019	0.019	1	0.040	1.037	0.1	0.110	0.216
10	0.104	0.113	0.033	0.033	2	0.083	1.078	0.2	0.240	0.349
20	0.188	0.188	0.058	0.057	3	0.126	1.120	0.4	0.480	0.585
30	0.267	0.281	0.083	0.084	4	0.167	1.161	0.6	0.705	0.824
40	0.337	0.370	0.106	0.105	5	0.205	1.196	1.0	1.157	1.274
Test	0.234	0.248	0.072	0.072	Test	0.103	1.098	Test	0.605	0.735
C	26.4	26.3	25.8	26.2	C	2.50	2.49	C	0.52	0.53
$\pm U$	±2.1	±1.2	±1.2	±1.2	$\pm U$	±0.08	±0.10	$\pm U$	±0.02	±0.02

Table 2 Acceptance criteria

Kind of analysis	Response	Criterion A		Criterion B			
		$\lvert C_{new}-C_{pre}\rvert$	$(U^2_{new}+U^2_{pre})^{1/2}$	$\lvert C_{pre}-C_{tr}\rvert$	$(U^2_{pre}+U^2_{tr})^{1/2}$	$\lvert C_{new}-C_{tr}\rvert$	$(U^2_{new}+U^2_{tr})^{1/2}$
Graphite furnace	Peak height	0.1	2.5	1.4	2.2	1.3	1.3
	Peak area	0.4	1.8	0.8	1.3	1.2	1.3
Flame	Absorbance	0.01	0.13	0.00	0.09	0.01	0.11
	Emission	0.01	0.03	0.02	0.03	0.03	0.03

Conclusions

1. The approach to the validation of a computer program as a validation of a component of the analytical method (including this program as "matreshka") is a productive one because of the use of the final results of the program application, i.e. the analytical results, for the validation conclusion.

2. The performance characteristics of the program (validation parameters) should be chosen from corresponding characteristics of the method and supplemented, if it is needed, with other ones specific for the program.

3. The validated program GIRAF corresponds to the requirements of international guides to quality in analytical chemistry [10, 11] and may be used for graphite furnace analysis as well as for flame absorption and emission analysis with the Perkin-Elmer 5000 spectrophotometer.

Acknowledgement The authors express their gratitude to Prof. E. Schoenberger for helpful discussions.

References

1. Green JM (1996) Anal Chem 68:305A–309A
2. Bedson P, Sargent M (1996) Accred Qual Assur 1:265–274
3. Huber L (1996) Accred Qual Assur 1:24–34
4. Burgess C (1996) VAM Bulletin 15:18–20
5. Huber L, Thomas M (1995) LG-GC 13:466–473
6. Huber L (1996) LG-GC 14:502–508
7. Samways K, Sattler L, Wherry B (1996) Pharm Technol 20:45–49
8. Kuselman I, Shenhar A (1997) Accred Qual Assur 2 (in press)
9. Guidelines for the Determination of Recalibration Intervals of Measuring Equipment Used in Testing Laboratories (1984) International Document No. 10 OIML, 1st edn. Bureau International de Métrologie Légale, Paris
10. Accreditation for Chemical Laboratories. Guidance on the Interpretation of the EN 45000 Series of Standards and ISO/IEC Guide 25 (1993) WELAC Guidance Document No. WGD2/EURACHEM Guidance Document No. 1, 1st edn. Available from local EURACHEM offices
11. International Guide to Quality in Analytical Chemistry. An Aid to Accreditation (1995) CITAC Guide 1, 1st edn. ISBN 0-948926-09-0, Teddington
12. Quantifying Uncertainty in Analytical Measurement (1995) EURACHEM, 1st edn. ISBN 0-948926-08-2, Teddington

Accred Qual Assur (1997) 2:275–284

Gordon A. Ross

Instrumental validation in capillary electrophoresis and checkpoints for method validation

Received: 15 January 1997
Accepted: 10 February 1997

G. A. Ross (✉)
Hewlett-Packard GmbH,
76337 Waldbronn,
Germany
Tel.: +49-7243-602663
Fax: +49-7243-602-149/501
email: gordon_ross@hp.com

Abstract Capillary electrophoresis (CE) is increasingly being used in regulated and testing environments which demand validation. The design, development and production of CE instrumentation should be governed by qualifications which ensure the quality of the finished product. The vendor should therefore provide guidelines and procedures which assist the user in ensuring the adequate operation of the instrumentation and especially in designing installation qualification (IQ) and operational qualification/performance verification (OQ/PV) procedures. OQ/PV should test those functions of an instrument which directly affect the CE analysis, i.e. voltage, temperature, injection precision and detector function. In validation of CE methods care should be taken that those aspects which directly affect the precision of peak parameters are appreciated. The relationship between CE instrumentation, chemistry and validation parameters is discussed and guidelines are presented for definition of a CE method for submission to regulatory authorities.

Key words Instrument · Method · Validation · Capillary · Electrophoresis

Abbreviations *CZE* capillary zone electrophoresis · *MECC or MEKC* micellar electrokinetic chromatography · *CGE* capillary gel electrophoresis · *CIEF* capillary isoelectric focusing · *CEC* capillary electrochromatography · *DAD* Diode-array detection

Introduction

Capillary electrophoresis (CE) is being used increasingly for quantitative analyses in regulated environments, testing laboratories and food and environmental analyses [1]. With the maturing of the technique, instrumental automation and increasing familiarity of workers with its operation, an increasing number of reports show that acceptable precision can be readily achieved, and rugged validated methods can be developed [2–9].

Validation of any instrumental analysis demands the validation of the component parts of the analysis and their combined function. This entails validation of the instrumentation, the analytical method developed on that instrumentation and their suitability for routinely performing the intended analysis. This report is intended to highlight instrumental and methodological validation issues specifically applied to CE and to discuss the relationship between vendors and users of CE instrumentation in environments demanding validation of such instrumental analysis.

Validation of capillary electrophoresis instrumentation

Comprehensive validation of any instrumental method of analysis is not achieved solely in the users analytical laboratory. The roles of the vendor and end user of the instrumentation are quite different. However, these roles do overlap significantly at the point of instrumental validation, and there are areas where the vendors contribution can be an invaluable aid to the user.

The vendor develops and delivers the instrumentation, while the user sets criteria and specifies tests to determine whether the instrumentation is suitable for the intended use. The differing responsibilities for setting and assessing such qualification and their overlap are shown in Fig. 1. In order to supply reliable instrumentation, its development, manufacture and the manufacturing process itself must be qualified. The process begins with the manufacture of instrumentation by vendors operating to ISO 9001 standards. For GLP purposes, any equipment includes the instrument (hardware) and any computer-operating system used with it (software). The process of validation may be viewed as three qualification phases. First is the design qualification (DQ) phase, where the instrument is defined, designed and developed, and second the installation qualification/operational qualification (IQ/OQ) phase where the instrument is installed and calibrated. Operational qualification is also referred to as performance verification. Last is the performance qualification (PQ) phase where the suitability of the analytical method for its intended use is monitored.

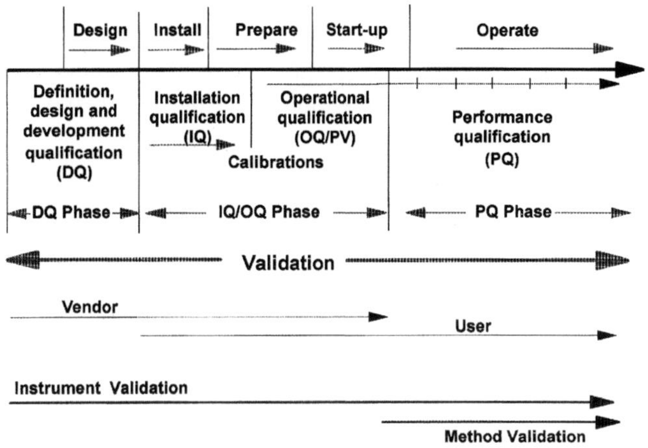

Fig. 1 The relationship between vendor and user during the qualification stages of instrumental and method validation

Design qualification (DQ)

Hardware

Instrumental specifications set by the vendor may be quite different from those required by the user. Vendor specifications, while helpful in indicating to the user whether an instrument is suitable for the analysis, are actually intended to indicate whether the instrument conforms to the definition of the instrument during its design, development and manufacture. The vendor has sole responsibility during the DQ phase of an instrument's lifecycle (Fig. 1). It is at the end of this DQ phase that the instrument is tested for conformity to the specifications set during this phase. These are the instrumental specifications. The vendor should therefore supply some form of qualification to indicate that the instrument conforms to the instrumental design specifications, although, as already stated, these can be quite different from the user's specifications. Manufacturers are often best placed to provide guidelines and procedures for operation and maintenance of the instrument and should work closely with the user to determine procedures such that the equipment's operation can be validated. However, GLP/GMP-bound end users should implement appropriate inspection and testing programs to test instruments for conformity to their own specifications, which dictated the choice of the instrument purchased in the first instance. The vendor can assist in this process and reduce the burden on the user if validation certificates can be provided with each instrument upon delivery. In this case, the need for the user to perform these tests is removed. A *declaration of conformity* from the vendor certifies the type of tests which the instrument has undergone.

Software

The majority of analytical processes in regulated environments rely on computer control for automation and for data capture and evaluation. It is therefore important that not only is the instrument validated but also the computer and operational software. Validation of software is more complex than hardware validation, especially since source codes for many analytical processes can run to several hundred thousand lines. In this case the responsibility for validation still remains with the user, who should have the assurance that the computer system has been validated during and at the end of the DQ phase. The vendor should provide documentation with a declaration that the software has been validated to these or similar guidelines to ensure software quality standards. Further, the vendor should provide access to the source codes if required by the user.

Installation qualification/operational qualification (IQ/OQ)

Installation qualification (IQ)

Installation qualification (IQ) for instrumentation is defined by the vendor. IQ provides documentation which shows that the instrumentation was complete upon delivery, correctly installed and functional. The IQ is performed by the vendor and should include a single test run the results of which should be stored with the IQ documentation.

Operational qualification/performance verification (OQ/PV)

Manufacturers can greatly assist the user by providing operational qualification/performance verification (OQ/PV) procedures to be performed on site to check the instruments performance. When an OQ/PV test procedure is developed for an instrument it is important to identify and test those functions of the instrument which have a direct effect on its ability to perform the analyses it was intended to undertake. This is sometimes confused with testing the instrument for conformity with instrumental specifications, but as previously stated the instrumental specifications are linked to the DQ phase and not subsequent use. The instrumental functions tested in OQ/PV are those which affect the qualitative and quantitative data required from the analysis.

The OQ/PV should be performed on installation, after extended use (e.g. at least once per year) and after major repair. This may also be extended to performing an OQ/PV before starting to validate or optimise a method on that instrumentation. OQ/PV for a capillary electrophoresis instrument should therefore be directed towards those specific instrumental functions which can affect the CE analysis.

A CE instrument comprises a high-voltage source and electrodes, containers for buffers, provision for holding and injecting samples, a thermostatted capillary housing and a detector. Most CE instruments use UV-vis absorbance detection, sometimes with spectral analysis capabilities. In instrumental separation sciences quantitative data is derived from the peak elution/migration time and peak dimensions; therefore any instrumental function which can affect these must be tested.

Parameters for testing in capillary electrophoresis OQ/PV

The parameters to be tested in OQ/PV for capillary electrophoresis are temperature stability, voltage stability, injection precision, detector performance (noise, drift and wavelength accuracy), and integrator functionality. Additionally, the migration time of a test analyte, using a well-defined method, may be monitored as an indication of the overall function of the instrument. Where the test procedure is performed using a test analyte and method, the usual validation parameters of the method should be determined. In all cases, care should be taken to develop a test procedure which directly tests the parameter in question and is not inferred from nor dependent on other instrumental or chemical parameters. The parameters to be tested are discussed in detail below.

Temperature stability

The capillary temperature can affect both peak migration time and peak area (Fig. 2a, b). The effects are due to temperature-mediated viscosity changes in the buffer. The electrophoretic mobility of analyte (μ_e) is proportional to its charge (q) to mass ratio, and inversely

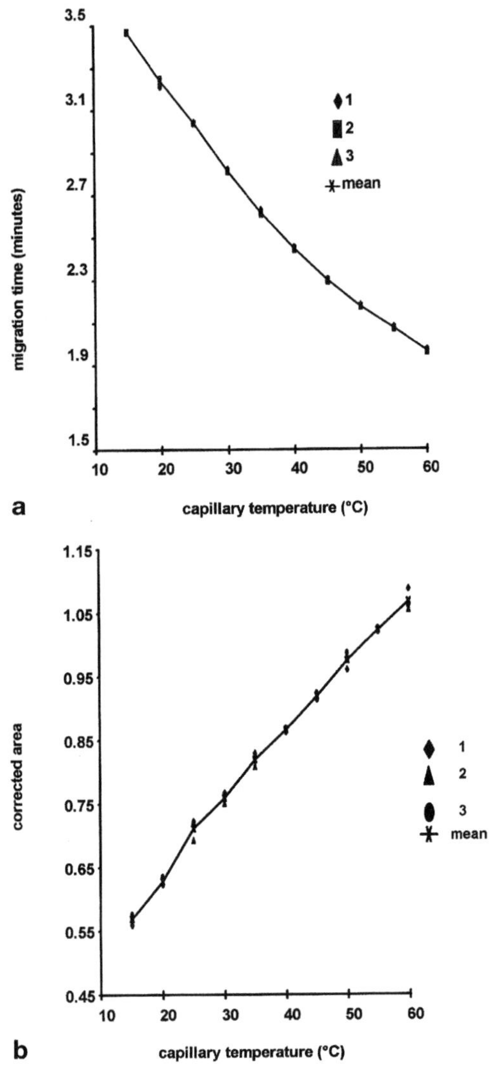

Fig. 2 The effect of capillary temperature on (**a**) peak migration time and (**b**) amount injected by pressure. Conditions: buffer 20 mM borate pH 9.2, sample 0.1 mM *p*-hydroxyacetophenone, capillary 48 cm (40 cm eff) × 50 μm, detection 192 nm, injection 250 mbar/s, voltage 30 kV

related to the viscosity (η) of the medium through which it travels (Eq. 1).

$$\mu_e = q/6\pi r\eta \quad (1)$$

Therefore, as the viscosity decreases the mobility increases and the migration time gets faster. The volume injected via hydrodynamic injection (pressure or vacuum) is also inversely related to the buffer viscosity through the Hagen-Poiseuille equation (Eq. 2).

$$\text{volume injected} = \Delta P d^4 \pi t / 128 \eta L \quad (2)$$

Therefore, as the viscosity of the buffer in the capillary decreases, a larger volume of sample is moved into the capillary for the same pressure differential.

The stability of the temperature setting is of more importance to the long-term reproducibility of an assay than to its absolute accuracy, although this is important for method transfer. Capillary temperature can also affect the selectivity in MECC or CEC analyses and also the resolution of a separation. Direct recording of the temperature of the thermostatting medium (air or liquid) is of more use than using inferred measurements, e.g. current or migration times, since these are also dependent upon non-instrumental parameters, e.g. buffer ionic strength and capillary history. Most CE instruments will provide a record of the thermostat temperature derived from probes which are part of the temperature control mechanism. If this is not possible, then an external temperature probe should be used. Typically the capillary temperature should be stable to within 0.1 °C and accurate to within ±1 °C.

Voltage stability

The applied voltage (V) divided by the capillary total length (L) is the field strength (E), which affects the peak migration time (mt) through its direct relationship to the analyte's apparent mobility (μ_{app}), which is the sum of the analyte's electrophoretic mobility (μ_e) and electroosmotic mobility (μ_{eo}):

$$\text{migration time} \propto (\mu_e + \mu_{eo}) V/L \text{ or } \mu_{app} E$$

Reproducible migration times are therefore instrumentally dependent on a reproducible field strength. Using a fixed capillary length, the voltage may be used as a direct indication of the stability and accuracy of the field strength. Directly recording the applied voltage is of more benefit than using inferred measurements such as migration time or current since these are dependent on non-instrumental chemical parameters e.g. buffer concentration. Where the instrument provides a record of the voltage applied, this may be used, but the vendor should indicate the source of the voltage measurement. Typically voltage supplies should be stable to within ±0.1% and accurate to within ±1% of the set voltage.

Injector precision

The ability of CE instrumentation to repeatedly inject a fixed volume of sample is fundamental to achieving a precise quantitative assay. Injector precision may be determined using a defined method (chemical and instrumental parameters), in conjunction with a test sample, by examining the precision of the reported corrected peak area (peak area/migration time). It is important to use the corrected peak area since this will remove minor fluctuations in migration time, due to chemistry, from the analysis of injector precision. While it is useful for the sample to be traced to national or international standards, this is not mandatory provided that the supplier operates to quality standards such as ISO 9001. The sample concentration should be such that it falls within the linear range of detection for that sample with that method and should provide sufficient signal-to-noise so that injector precision is not coloured by integration fluctuations. Optimally this should provide a peak with a signal-to-noise ratio of around 500. Other instrumental functions must be verified prior to performing this test where they can affect the amount injected or the reported peak area, i.e. detector linearity, capillary temperature and integrator function. Most CE instruments inject sample by applying a fixed or variable differential pressure for a fixed time. Their ability to precisely reproduce the injection is compromised for very short (ca. 1 s) injection times, therefore time of injection should be greater than 3–4 s. Simply dipping a capillary into a sample will result in a small volume being injected. In order to minimise the effects of this extraneous injection on the analysis of injector precision, a reasonable injection volume should be used to provide data. An injected volume of around 1–5% of the total capillary volume is recommended.

Detector performance

The majority of CE instruments are equipped with UV-vis absorption detection. The most important characteristics of UV-vis detectors which should be tested are the noise and drift. These should first be determined in the *absence* of a capillary using the ASTM method so that the detector is appraised directly and the measurements are not affected by the capillary. Fixed and variable wavelength detectors should be verified for accuracy of the detection wavelength. DAD and rapid-scanning detectors should be verified for wavelength accuracy over the defined spectral range by checking for accuracy at a minimum of three wavelengths which cover this range. This may be performed using an internal filter (e.g. holmium oxide) or using an analyte with a well-characterised spectrum. Detector linearity should be determined for the test analyte to ensure that the concentration used for determination of injector reproducibility is within the linear range of that analyte on the instrument being tested. Therefore the detector should be subjected to different tests: (1) noise and drift without a capillary present, (2) wavelength accuracy without a capillary present, (3) wavelength accuracy using a test analyte, and (4) linearity of response vs test analyte concentration over the range which covers the concentration used for the injector precision test.

Integrator functionality

The long-term stability of the integrator function can be determined using a test electropherogram. Integration parameters such as migration time, peak area, corrected area and baseline may be used to determine and monitor the functionality of the integration software. It is useful if the test chromatogram is appropriate to the analysis being undertaken by the instrument, and therefore the user should have the opportunity to define the test electropherogram.

Instrument-specific functions

Should a manufacturer supply an instrument with a specialised or uncommon feature which can affect the performance of an assay, e.g. buffer replenishment, then its functionality should also be assessed, especially if the function is part of the user's specification for suitability of the instrument to perform the intended analysis.

Performance criteria and limits

Setting the limits for determining whether the instrument is still within acceptable performance criteria should be entirely the responsibility of the end user. The vendor can assist in providing guidelines as to how the instrument should function on delivery, but the operational qualification is intended to determine whether the instrument is still operating to within the specifications which the user defined upon purchase. For example, if the intended analysis is quantitation of a main component in a drug formulation, then detector noise which affects sensitivity may be of lesser importance than injection precision. Conversely, where the instrument is used for impurity determination, especially where the reported data is (area/area) %, then sensitivity and therefore detector noise may be of more importance than injector precision. If limits are indicated by the manufacturer, then these may be used directly if appropriate, but the user should be aware that the decision is his or hers. The limits for an OQ/PV should not automatically be defined from the instruments specifications.

Checkpoints for validation of capillary electrophoresis methods

Analytical method validation has developed within the pharmaceutical industry over the years in order to produce an assurance of the capabilities of an analytical method. A recent text on validation of analytical techniques has been published by the international Conference on Harmonisation (ICH) [19]. This discusses the four most common analytical procedures: (1) identification test, (2) quantitative measurements for content of impurities, (3) limit test for the control of impurities and (4) quantitative measurement of the active moiety in samples of drug substance or drug product or other selected components of the drug product. As in any analytical method, the characteristics of the assay are determined and used to provide quantitative data which demonstrate the analytical validation. The reported validation data for CE are identical to those produced by an LC or GC method [11] and are derived from the same parameters, i.e. peak time and response. Those validation parameters featured by the ICH (Table 1) are derived from the peak data generated by the method. Table 1 also indicates those aspects of a CE method (instrumentation and chemistry), peculiar to the technique, which can affect the peak data and highlights factors which can assist the user in demonstrating the validation parameters.

Selectivity

Selectivity is demonstrated by identifying a peak as a single analyte. This may be achieved by characterising the peak by its migration time provided that it can be demonstrated that no other components are co-migrating with that peak. This may be demonstrated by co-injection of the sample with a reference standard. Spectral analysis software capable of providing UV-vis absorbance spectra and determining peak purity can be very useful in demonstrating the selectivity of an assay, especially in identification tests.

Peak purity

Using DAD or scanning detectors it is possible to determine the purity of a peak by spectral peak purity analysis. Peak homogeneity can be demonstrated by a variety of methods. However, the most commonly used technique normalises and compares UV spectra from various peak sections.

Table 1 The relationship between instrument and chemistry in CE analyses: the reported data and validation parameters

Validation parameters	Reported data	Instrument	Chemistry
Selectivity (Specificity)	Migration time (preferably in conjunction with peak purity)	Detection wavelength Peak spectra	Buffer pH Buffer components, additives and their concentrations
Precision (Accuracy) Stability Ruggedness Robustness	Migration time precision	Capillary temperature Applied voltage	Buffer pH, capillary treatment, capillary temperature, buffer ionic strength, organic modifiers
	Peak area precision	Capillary temperature Injection mechanism: pressure/vacuum applied Time of applied pressure/vacuum Height of vial and transition time	
Linearity Range	Linearity	Linear range of detector	
LOD LOQ	Sensitivity	Detector noise Data analysis Amount injected	

Peak identity

This may be determined by acquiring the UV spectra of a peak and comparing this with a known standard. The process may be optimised by constructing a spectral library and comparing the analyte spectra with that of a known substance in the spectral library. Using suitable software, this may be performed automatically and with a high degree of accuracy. Such spectral analysis of analytes is one of the main techniques used for peak identification in pharmaceutical, food and environmental analysis. Peak identity may also be determined from spiking the sample with a pure preparation of the presumed analyte and observing increase in peak height or area. Care should be taken that the increases in peak areas do not result from erroneous analyte identification and co-migration of two different species. Increases in peak widths or changes in the peak shape of the spike compared with the sample may indicate a co-migration of two peaks, although in CE peak shapes are frequently concentration dependent. Peak purity analysis on the spiked sample can be used to confirm the presence of a single analyte.

Migration time precision

Migration time precision may be used for peak identification and is important in assessing the overall performance of a method, since many other parameters impinge on its reproducibility. Recent studies have demonstrated that not only is CE capable of exhibiting good precision in a quantitative assay [3, 6, 7, 12] but that this precision can be preserved when methods are transferred between laboratories and between instruments [4, 5].

As we have seen from Eq. 1, the mobility is dependent upon the buffer viscosity, assuming that the buffer make-up is constant and its pH is stable. Therefore, from an instrumental point of view, the electric field and the capillary temperature are the two main parameters which must be stabilised to effect reproducible migration time. Where a buffer is selected for properties other than its buffering capacity, e.g. indirect detection techniques, it may be necessary to replace the buffer frequently in order to ensure reproducible migration times.

Peak area precision

Peak area precision is of great importance in the development of a quantitative assay. The reproducibility of the peak area is dependent upon a number of parameters, all of them instrumentally derived. Chemistry has little impact upon the quantitative reproducibility of an assay except where electrolytic changes in buffer pH can affect the absorption spectra of an analyte or where wall adsorption of the analyte can occur [17]. Most commercial instrumentation allows the use of electrokinetic and pressure injection modes for introduction of sample.

Electrokinetic injection with positive polarity introduces a bias towards the more cationic components of a sample [18]. Electrokinetic injection is, however, often necessary when performing CGE separations, since pressure injection may disturb the gel. In CGE analysis of nucleic acids and SDS proteins, smaller analytes will be preferentially loaded since they migrate faster into the gel. Other components of the sample matrix can also affect the amount of sample/analyte injected using electrokinetic loading. For quantitative analysis, the fluctuations and dependence of electrokinetic injection techniques on an analyte's charge and sample matrix generally preclude it use in quantitative assays. Injection by hydrodynamic displacement (pressure or vacuum or hydrostatic displacement) is a more useful technique and is more appropriate to quantitative analysis.

The precision of the amount injected by pressure is also dependent on the reproducibility of other instrumental parameters. The injected amount depends on precision of the applied pressure, the time control and also the rise and fall times of the applied pressure. In order to optimise the precision of the injected amount it may be necessary to use post-sample plugs of buffers or voltage ramps to minimise thermal expansion of the buffer plug and sample loss. Precise sample injection also depends on the viscosity of both the sample and the buffer, as indicated by Eq. 2. Since viscosity is temperature dependent, thermostatting the sample tray or operating in an environment where ambient temperature is controlled will eliminate the effects of temperature-induced sample viscosity variations. However, even with constant temperature in the sample tray, temperature-induced variations in the viscosity of the buffer in the capillary can also affect the amount injected (Fig. 2b). In this case a stable capillary temperature is important. Where a viscous sample is being analyses, care should be taken that calibration standards are of similar viscosity or, ideally, constructed using blank sample matrix [8]. Although the amount injected is determined by these instrumental factors, the reported area of the integrated peak is dependent upon the functioning of the integration software. Reproducibility of peak area generally degenerates as the concentration decreases (Fig. 4). This is in part due to integration effects where the integrator can have some difficulty in deciding the start and end of a peak which has a height of only two to three times that of the baseline noise.

The geometry of the capillary end plays an important part in ensuring reproducible peak shape. If the end of the capillary is ragged or broken, this can lead to carry-over effects [14]. Figure 4 shows the effects of capillary end geometry on the peak shape of thiourea analysed using CEC. The rise in baseline after the peak caused by a ragged capillary end can lead to errors in integration and therefore irreproducibility of the reported peak area. Ideally the capillary end should be flat and a short section of polyimide should be removed from it. Capillaries may be cut using a variety of means, e.g. a sapphire cutting tool or a ceramic stone. After the capillary is scribed, the break should be made by pulling the capillary apart and not by bending it. Alternatively, precision cut capillaries may be purchased.

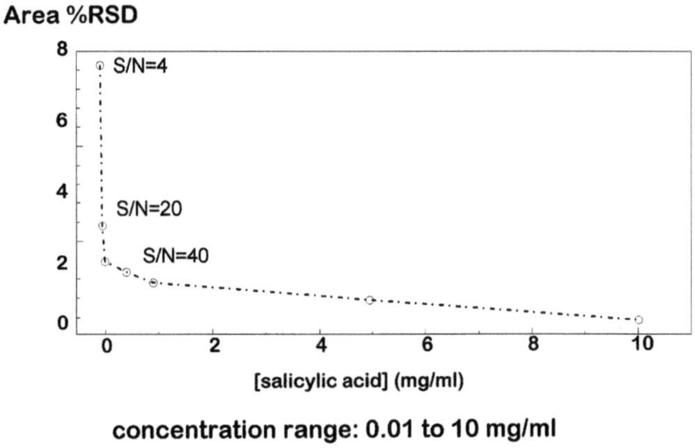

Fig. 3 The relationship between signal-to-noise ratio and precision of reported peak area

Detector linearity, limit of quantitation and limit of detection

The linearity of an analytical method is its ability to produce test results which are directly, or by means of a well-defined mathematical transformation, proportional to the concentrations of analytes within a given range. Usually it is important to demonstrate that a linear relationship exists between detector response and

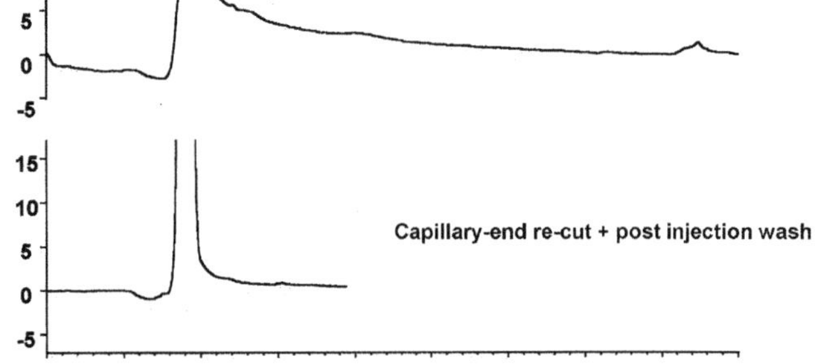

Fig. 4 The effects of capillary end geometry and capillary end washes on peak shape

analyte concentration. Linearity should be determined by injection of at least ten standards of varying concentrations which cover 50–150% of the expected working range of the assay. In CE, using direct detection methods, a linear range covering three to four orders of magnitude is obtainable. However, in indirect detection modes, linear ranges of two orders of magnitude are common. A linear regression analysis applied to the results should have an intercept not significantly different from zero. Errors associated with each concentration level should be determined by multiple injection at each concentration level. Linear ranges of three to four orders of magnitude are of great utility in the determination of low (<0.1%) impurity levels. The limits of detection of an assay using UV detection can be determined as that signal which gives an S/N ratio of 3. Therefore, with a knowledge of the molar absorptivity of the analyte, this can be determined purely from a knowledge of the noise of the detector. A caveat to this is that this noise should be the noise of the "system", i.e. the detector noise at the detection wavelength during a run.

The limit of detection (LOD) is that concentration which can be differentiated from a blank, while the limit of quantitation (LOQ) is the lowest concentration which can be determined with an acceptable accuracy and precision. It is important to realise that with most instruments, when a capillary is replaced, in most cases the detector flow cell is also replaced since detection takes place on-column. In this case, although the entire method need not be re-validated, the limits of detection and limits of quantitation, if these are appropriate to the method validation being undertaken, should be redetermined. Minor variations in capillary alignment can be sufficient to produce significant changes in detector response and sensitivity. This may not be necessary when using a decoupled detection cell.

In contrast to OQ/PV testing, the detector baseline noise should be determined using the analytical buffer with the operating voltage applied. For fixed-wavelength detectors, the wavelength used for detection and quantification should be chosen such that this lies on a plateau of UV absorption for both analyte and internal standard. With the use of a DAD, where multiple wavelengths can be monitored, the sample detection wavelength may be different from that used for the internal standard. This provides a greater flexibility in choice of internal standard, if used, and in choice of detection wavelength to provide maximum absorption.

Definition of a CE method for submission to a regulatory agency

When detailing a method, care should be taken that the appropriate information is provided such that not only can the method be reproduced on a similar instrument by a different operator, but also there is sufficient detail to facilitate method transfer between instruments from different manufacturers. These details should also be included when submitting to a regulatory agency. Ideally the method description should take a standard form. A checklist of method parameters for submission of a CE method is shown in Table 2.

All materials used to construct the buffer should be made from electrophoresis grade materials. It is of

Table 2 Checklist for detailing a CE method for submission to regulatory authorities

Parameter	Required information
Purpose	Nature of the analysis Reasons for performing the analysis Analyte being determined.
Instrument	Instrument manufacturer Model
CE mode	Separation mechanism CZE, MECC, CGE, CIEF, CEC
Buffer	Buffer constituents, concentrations and pH Weight of each component identified by chemical formula as well as common names Volume and concentration of acid or base used to adjust the pH Total volume of solvent (e.g. water) Ready-made buffers: source, batch number and specifications.
Capillary	Total capillary length (inlet to outlet, cm) Effective or detection length (inlet to detection point, cm) Internal diameter (including extended lightpaths if used) Type: bare fused silica coated (type of coating and vendor) packed (packing material and vendor)
Injection	Injection mechanism: Electrokinetic: applied voltage and time of application Hydrodynamic: units of pressure or vacuum applied or height of displacement and time of application
Voltage	Applied voltage (field strength i.e. volts per cm of capillary) Expected current (i)
Temperature	Capillary thermostat temperature Sample tray temperature
Detection	Type of detector (e.g. DAD, fixed wavelength etc) Detection wavelength, band width, rise time and sampling rate Reference wavelength if using DAD
Capillary/ /reatment	Prior to first use Pre-analysis Inter-analysis Storage Indicate conditioning solutions, flush pressure and time of treatment

some importance to note that where surfactants or organic additives are used the buffer should have its pH adjusted before their addition, and this should be noted. The construction of the buffer should take the form of a standard operating procedure (SOP). Where ready-made buffers are used for the analysis, the source, batch number and specifications of the buffer should be indicated. Pre-made buffers should be supplied by the vendor with appropriate documentation indicating assay information and verification of purity. Since instruments from different manufacturers use different units of pressure or vacuum for sample injection it is of *specific* importance to specify these. In particular, for method transfer, the pressure units should be translated into the pressure units of the instrument the method is being transferred to. Similarly, in instruments of different manufacture, the standard capillary dimensions, i.e. total and effective lengths, may vary. It is also very useful to indicate the expected current with the indicated buffer, field strength and temperature settings since this may highlight errors in any of these parameters.

The capillary pre-treatments should be documented as part of the method. It is generally sufficient to indicate the time of treatment and the conditioning solutions used. However, with transfer of a method between instruments the number of column volumes should be indicated since different instruments use different pressures to flush capillaries. The reasons for detailing capillary pre-treatments lie in the influence on analyte mobility of the electroosmotic flow (EOF) which is ubiquitous in bare fused silica capillaries. Since the mobility of an analyte is composed of its own mobility plus that of the EOF then this must also be stable in order to produce reproducible migration times. Stability of the EOF can be achieved by appropriate capillary pre-conditioning techniques which should be suited to the analysis being undertaken, e.g. preconditioning with 10% v/v phosphoric acid prior to peptide separations using phosphate buffer at low pH or cleaning the capillary between runs with 1 N NaOH when analysing samples where either analyte or sample matrix exhibits some wall adhesion. This is necessary because of the hysteresis of the relationship between pH and EOF when increasing or decreasing the buffer pH [13–15].

There are various types of capillary pre-treatment. One is the prior-to-first-use treatment; other treatments include preparative conditioning performed either daily or after extended use and capillary treatment between analyses. Conditioning between analyses may or may not be necessary, although minimally this should include a buffer flush to eliminate the possibility of inter-analysis carry-over. Finally details of capillary treatment prior to storage should be documented if the analytical method is intermittently used.

Table 3 shows the EOF associated with various batches of silica with and without pre-treatment. Preconditioning the capillary with the indicated wash regimen serves to improve reproducibility of the EOF in all cases, although there is still a marked variability between silica batches. Sufficient time should be allowed for the capillary to equilibrate with the run buffer. For conditioning between analyses, flushing with buffer only may be sufficient. More elaborate techniques such as applying a short duration of voltage after flushing with run buffer has been found to improve migration time reproducibility with some analyses [16]. When using coated capillaries it is usually not necessary to use any wash technique and is sometimes inadvisable. However, buffer should be replaced between runs to eliminate the risk of fouling or carry-over, and storage conditions where appropriate should be detailed. In all cases one capillary should be used for one method only.

Table 3 EOF variations with silica batch and pre-treatment

Silica batch	Not pre-treated		Pre-treated[a]	
	mean* EOFmm/s	% RSD	mean* EOF mm/s	% RSD
QQT11A	0.2307	1.31	0.2837	0.86
QNRO1	0.2353	1.18	0.2828	0.78
MNZ04B	0.2303	5.95	0.2811	1.48
KZG01A	0.1886	23.97	0.2580	1.71
MSZ01	0.1538	11.32	0.2195	0.95
KYL12	0.2185	13.8	0.2517	0.42
KYL05	0.1406	19.82	0.2900	0.97
All data	0.1997	19.65	0.2667	9.47

* All $n=10$

[a] pre-treatment 5 min methanol, 5 min NaOH, 5 min water, 20 min run buffer (flush pressure 1 bar)

Conditions: capillary 48.5 cm (40 cm eff) × 50 μm id, buffer 50 mM phosphate pH 7.00, temperature 20 °C, injection 65 mbar/s, voltage 26 kV, sample 0.2% DMSO in water

Summary

The requirements for validation of CE instrumentation have been described. Although these are broadly similar to the requirements for validation of any instrumentation used in GLP/GMP-compliant laboratories, some aspects are peculiar to CE, especially in regard to OQ/PV testing. Those features of a CE instrument which should be tested in an OQ/PV include temperature and voltage stability, detector function and injector precision. These should be directly assessed and not inferred from other measurements dependent upon one or more of these parameters or upon chemistry. It is important that the contributions to the performance of an analysis are identified as instrumental and/or chemical. Capillary electroseparation methods are fully capable of being validated similarly to LC and GC methods with identical validation criteria. When defining the parameters of a CE method, all operational parameters should be noted, including accurate instructions for construction of the buffer and also the capillary pre-treatment used. Guidelines for detailing a method for submission to regulatory agencies must include sufficient detail for reproducibility studies at all levels as well as for method transfer between instruments.

Acknowledgement The author would like to thank Ludwig Huber, Hewlett-Packard, Waldbronn for critical reading of the manuscript and productive discussions.

References

1. Altria KD, Kersey MT (1995) LC-GC 8 (4):201–208
2. Altria KD (1993) J Chromatogr 634:323–328
3. Altria KD (1993) J Chromatogr 646:245–257
4. Altria KD, Harden RC, Hart M, Hevizi J, Hailey PA, Makwana JV, Portsmouth MJ (1993) J Chromatogr 641:147–153
5. Altria KD, Clayton NG, Hart M, Harden RC, Hevizi J, Makwana JV, Portsmouth MJ (1994) Chromatographia 39:180–184
6. Prunonosa J, Obach R, Diez-Cascon A, Gouesclou L (1992) J Chromatogr 574:127–133
7. Thomas BR, Fang XG, Chen X, Tyrrell RJ, Ghodbane S (1994) J Chromatogr 657:383–394
8. Perrett D, Ross GA (1995) J Chromatogr 700:179–1865
9. Thomas BR, Ghodbane S (1993) J Liq Chromatogr 16:1983
10. Huber L (1994) Good Laboratory Practice and current good manufacturing practice. A primer. Hewlett-Packard publication number 12-5963=2115E
11. Altria KD, Rudd DR (1995) Chromatographia 41:325–331
12. Altria KD, Fabre H (1995) Chromatographia 40 (5/6):313–320
13. Lambert WJ, Middleton DL (1990) Anal Chem 62:1585–1587
14. Ross GA (1994) Biomedical applications of capillary electrophoresis including the analysis of structural haemoglobinopathies. PhD thesis, University of London, UK
15. Coufal P, Stulik K, Claessens HA, Cramers CA (1994) J High Res Chromatogr 17:325–334
16. Ross GA (1995) J Chromatogr 718:444–447
17. Soga T, Ross GA (1997) J Chromatogr (in press)
18. Huang XH, Gordon MJ, Zare RN (1987) Anal Chem 60:375–377
19. Pharmeuropa Vol. 5.4 (Dec. 1993) 343–345

Accred Qual Assur (1997) 2:360–366

Ludwig Huber
Herbert Wiederoder

Qualification and validation of software and computer systems in laboratories

Part 1: Validation during development

Received: 17 May 1997
Accepted: 30 June 1997

L. Huber (✉) · H. Wiederoder
Hewlett-Packard Waldbronn,
PO Box 1280,
D-76337 Waldbronn, Germany
Tel.: +49-7243-602 209;
Fax: +49-7802-981948;
e-mail: Ludwig_Huber@hp.com

Abstract Software and computer systems are tested during all development phases. The user requirements and functional specifications documents are reviewed by programmers and typical anticipated users. The design specifications are reviewed by peers in one to two day sessions and the source code is inspected by peers, if necessary. Finally, the function and performance of the system is tested by typical anticipated users outside the development department in a real laboratory environment. All development phases including test activities and the final release follow a well-documented procedure.

Key words Validation · Qualification · Computers · Software · Analytical · Laboratories

Introduction

Proper functioning and performance of equipment play a major role in obtaining consistency, reliability and accuracy of analytical data. Therefore equipment should be properly selected, designed, installed, and operated, and the correct function and performance should be verified before and during operation. This holds for equipment hardware, computer hardware, hardware interfaces, and software and computer systems. Qualification of equipment hardware is well established and has been described by several authors [1–4], and typically users in analytical laboratories are quite familiar with testing equipment for hardware specifications.

Software and computer system validation differ from hardware validation in that it is harder to specify absolute performance criteria and functional specifications for software and to define tests and acceptance criteria. There are many publications that deal with the validation of software and computer systems in one way or another. However, either they are not specific for the requirements of analytical laboratories or they are not specific to computers in the laboratory.

This series of articles should fill this gap. It was felt that the subject is so complex that it would not fit into a single article. Therefore, four articles are currently planned. The first deals with validation and qualification during development. The second deals with the tasks that should be performed during installation and prior to operation. Article number three covers the validation tasks during routine use. While the first three articles deal with the validation of new systems, article number four gives recommendations on how to retrospectively evaluate and validate existing systems.

This first article describes the validation and qualification of computer systems such as those for instrument control and data evaluation during development. Development validation of computer systems purchased from a vendor typically is done at the vendor's site, and even though most computer systems and software in analytical laboratories are purchased from vendors it was felt that such an article makes sense for users of such systems for two reasons:

1. From a regulatory point of view, computer systems must be validated. The user has the ultimate responsibility for validation but he can delegate some parts, for example validation during development, to the vendor. If he does this, the user should have some proof that his software has been validated at the vendor's site. Using software development practices

at Hewlett-Packard Waldbronn as example, this article should help users to understand what should be done during development, and this information may be used to ask the right questions of the vendor.
2. Users of computer systems may develop software on their own either as a stand-alone package, such as a software for specific statistics, or as an add on to the standard software. This software should be validated and programmers should get ideas and advice on how to validate.

Because there is still a misunderstanding of terms such as validation, qualification and verification, these will be explained right at the beginning. It is also important to understand the terms computer system and computerized systems and the different types of software loaded on a computer, but other publications should be consulted for this type of information.

Readers of the article, who have to comply with regulations should know about and understand them. Relevant information can be found in existing literature [7, 8].

Definitions

The terms validation, verification and qualification are frequently used interchangeably. Validation and verification have been defined by EN ISO 8402:1995 [6].

Verification: Confirmation by examination and provision of objective evidence that *the requirements* have been fulfilled.

Validation: Confirmation by examination and provision of objective evidence that *the particular requirements for a specific intended use* are fulfilled.

Even though the definitions look similar, there is a distinct difference. While verification is of general nature, validation refers to 'specific intended use'. In this sense a computer system that is developed for multiple users with multiple applications is verified rather that validated at the vendor's site. When the system is installed at the user's site for a specific task and the system is tested to meet the previously specified requirements, this process is defined as validation. If the system is intended to be used for different applications and more generic tests are done in the sense of EN ISO 8402:1995, this process again is called verification.

In practice, vendors of computer systems and user firms use the terms differently. The pharmaceutical industry uses the terms validation in the sense of EN ISO 8402:1995, and the term verification is hardly known. Instead of this the term "qualification" is used.

The entire qualification process is broken down into

- Design qualification (DQ) for setting functional and performance specification (operational specification)
- Installation qualification (IQ) for performing and documenting the installation in the selected user's environment
- Operational qualification (OQ) for testing the equipment to ensure that it meets the previously defined functional and performance specifications
- Performance qualification (PQ) for testing that the system performs as intended for the selected application.

OQ is similar to performance verification. PQ is most similar to validation as defined by EN ISO 8402:1995, because PQ always includes the user's equipment and method, and therefore is very specific.

Instrument vendors use the term qualification for installation and both verification and qualification for testing the equipment hardware and software for documented specifications prior to operation. For software development typically the term validation is used, although the term verification would be more appropriate.

The confusion is made nearly complete by the introduction of new terms by official committees. For example, the OECD has used the term acceptance testing [9] which is a subset of OQ and should be performed before the full OQ as a criterion of whether the system would be accepted by the user's firm.

Software product life cycle

Software development often takes several years, and it is impossible to ensure a certain quality standard simply by testing the program at the end of its development process. Quality cannot be designed into the software after the code is written; it must be designed and programmed into software prior to and during its development phases by following written development standards, including the use of appropriate test plans and test methods.

The product life cycle approach as illustrated in Fig. 1 has been widely accepted for the validation of computerized systems during their entire life. The product life is divided into phases:

- Setting user requirements and functional specifications
- Design and implementation, with code generation and inspection
- Test of subsystems (build a system and test as a system)
- nstallation and qualification of system before it can be put into routine use
- Monitoring performance of system during its entire use
- Maintenance and recording history of changes

The Hewlett-Packard product life cycle starts with the proposal to initiate a new project. Proposals are

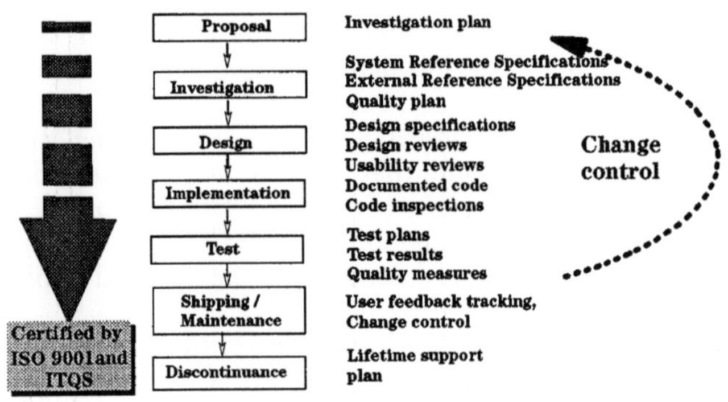

Fig. 1 The Hewlett-Packard CAG product life cycle model. ITQS: Information Technology Quality system

based on business and user needs. These describe how users fulfill these needs now and how the new software can provide better solutions. In the investigation phase, system reference specifications (user requirements) and external reference specifications (functional specifications) are developed and reviewed. In the design phase, the design specifications are developed and reviewed. The implementation phase includes writing the code and code inspections, if necessary. In the test phase, functional testing is performed with test cases for each function as specified in the external reference specifications document. After the product is shipped and used, feedback from users is recorded and documented, and changes to the software are made following documented change control procedure. Each phase is completed, reviewed and approved before the subsequent phase is started.

Checkpoint meetings

Each phase ends with a checkpoint meeting. This is prepared by the project leader and attended by all members of the project team and managers from different departments. Team members report on their activities during the phase and how they could meet the requirements as written in the development and validation plan. The team and management go through the checklist and discuss each point to determine whether and how the checkpoint items have been fulfilled. An action plan with people assignment is put together as part of the phase exit report for those items that are not yet closed. If all issues are resolved, the checklist is signed off by the management.

Proposal and investigation phases

The software life cycle starts with a requirements analysis and product definition. These define the requirements that the product must meet for functionality, compatibility with existing systems, usability, performance, reliability, supportability and security. The goal is to specify both the problem and the constraints upon the solution. Planning activities include project plans, budgets, schedules and validation, verification and testing. During this phase the project team is established, usually comprising representatives from system development, product marketing, product support, quality assurance, manufacturing and application chemists, who represent the users and are deeply involved in the development of a functional requirements specification and in the user interface prototyping. A project team leader is appointed to manage the project and a project notebook is created and maintained through the entire development phase.

Users from all application segments are interviewed by team members to discover their needs. Finally, a list with all proposed functional requirements is drawn up and evaluated by the project team. Usually the list is too long for all requirements to be implemented within a reasonable time-frame, so the requirements are prioritized into three categories, “Musts”, “Wants” and “Nice to haves”. “Must” requirements are considered to be those that are a prerequisite to the success of the software and are always included in the final specifications. “Wants” and “Nice to haves” are of lesser importance and are included only if their implementation does not appreciably delay the project.

The external reference specifications (ERS) document is developed. This includes an overview of the scope and benefits of the project and a detailed description of the complete product from a user’s perspective. The document is reviewed by the project team. It is the starting point for system design and also the basis for

- Design specifications document
- Functional testing
- The functional specifications document that is available for users to write their own requirement specifications and operational qualifications or acceptance testing
- The user documentation (e.g., user manual, on-line help).

Feasibility studies are done if necessary. The software engineering tools are determined and the software ‘make’ process is designed.

Deliverables for this phase include:

- System reference specifications (user requirement specifications)
- External reference specifications (functional specifications)
- Risk assessment
- Quality plan

Design phase

The goal here is to design a solution that satisfies 100% of the defined 'must' requirements and falls within the set constraints. Alternative solutions are formulated and analyzed, and the best are selected. Verification activities during the design phase include inspecting the design specifications for completeness, correctness, consistency and checking that the design directly correlates with the defined requirements. Thorough design inspections are of the utmost importance because correction of defects detected in this phase is much less costly than is the case when these are detected in a later life cycle phase.

Details on system screen designs, report layouts, data dictionary with data flow diagram, system configurations, system security, file design, system limitations and memory requirements are laid out by system developers and are usually formally inspected with members of the development team. Major outputs of this phase are the internal design documents and prototypes. The design documents are based on the ERS and can be used as a source for the technical support documentation.

In this phase, operators from different backgrounds test the user interface in a process called usability testing to determine the access to the intended function and their understanding of interaction concepts.

The design specifications as prepared by individual programmers are inspected by peers. This is done in a meeting organized and led by a specially trained moderator who may be from the development department or the quality assurance department. Programmers review the design specifications document for consistency and correctness. Any findings are discussed and recorded as shown in Fig. 2. The procedure is repeated until the team no longer finds any defects in the document. Deliverables for this phase include:

- Internal design document (design specifications)
- Reports on design inspections/reviews

Date:4/20/96
Moderator: Klaus Weber

- Iterative sessions
- With moderator

#	Page	Line	Description	Feedback from	Date fixed
1	1	6	Comma at the end should be in brackets	DH	6/4/96
2	1	13	Should be stated that '#' is needed	FH	6/4/96
3	2	2	Give an example what this is good for	NP	6/8/96
4	2	4	Is it possible to add start location of the dialog box	MZ	6/8/96

Fig. 2 Extract from a design review

- Usability test report
- GLP validation/documentation plan
- Chemical performance and application test plan
- QA plan update

Implementation phase

In the implementation phase, the detailed design is implemented in source code, following written and approved software coding standards, and results in a program that is ready to be tested. After certain groups of functions have been programmed, they are tested individually by the programmers before they are integrated into a larger unit or into the complete system. Verification includes code and internal documentation, test designs, and all activities that determine whether the specified requirements have been met. Concurrently with the implementation phase, system documentation, such as user manuals and the electronic help system, is prepared. Documentation also includes a description of the algorithms used by the program.

In this phase, the system also undergoes a rigorous usability evaluation with testers from different backgrounds. The goal is for an experienced user to be able to perform the basic functions without the need for formal instruction (the so-called plug-and-play approach).

Deliverables for this phase include:

- Source code with documentation
- Code inspection/walkthrough reports, where necessary
- Documentation of test cases in preparation for the test phase.

Testing

Thorough testing and verification are most important for any validation and qualification. For a software project, testing and verification are done throughout all life cycle phases. The goal is to detect errors, if any, as early as possible. Requirements specifications and the design are reviewed or inspected during the definition and design phases and the code is tested and may be formally inspected by the programmers during code implementation. Proper functioning of software together with the equipment hardware is verified in the test phase and during operation.

Types of testing

Software testing can be classified as being either structural (white box) or functional (black box) (Fig.3). Structural testing (white box) of software is the detailed

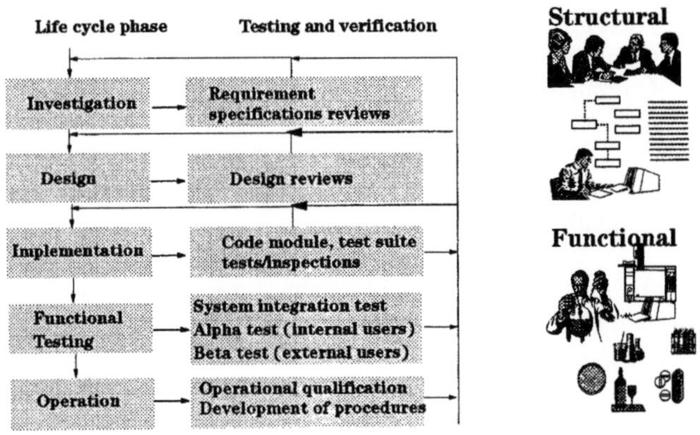

Fig. 3 Testing and verification are done throughout all life cycle phases

examination of the internal structure of code, including low- and high-level path analysis, branch flow analysis and inspection for adherence to software development procedures. It tests logical paths through the software by providing test cases that exercise specific sets of conditions. Besides the source code, other documentation, such as logic diagrams, branch flow analysis reports, description of modules, definition of all variables, specifications, and test suites of all inputs and outputs, are required.

Functional testing (black box) of software evaluates the outputs of a program compared to the expected output values for a range of input values. For a computer-controlled analytical system, functional testing should always include analytical hardware to verify proper parameter communication and data flow. Source code is not required, but a full set of system specifications and a description of functional routines, such as calibration algorithms, must be available.

Structural testing is done in the development department and starts during the implementation phase. Code modules are checked individually by the programmers and may be formally inspected by peers, if appropriate. Modules are tested by programmers with specific test suites and then linked together into a system and tested as a whole for proper functionality to ensure that designs are correctly implemented and the specified requirements satisfied.

Written in advance, the test plan defines all test procedures with their pass/fail criteria, expected test results, test tasks, test environment for equipment and computers, criteria for acceptance and release to manufacturing, and the persons responsible for conducting these tests. The test plan also specifies those functions excluded from testing, if any. Individual tasks cover functional testing, simulation of incomplete functions as integration proceeds, mathematical proof of results, records of discrepancies, classification of defects and corrective actions.

System requirement specifications	Requ 1	Requ 2	Requ 3
External reference specifications	Funct 1	Funct 2	Funct 3
Design review	x	x	x
Code inspection	x	x	x
Alpha-testing	x	x	x

Fig. 4 Test coverage matrix

A test coverage matrix (Fig. 4) is created which shows linkages of test cases to the design specifications and external (functional) and system (user) requirement specifications. This matrix ensures that all functions are tested through all phases.

α-Testing

After the programmers have completed the engineering tests, the system undergoes functional testing in typical operating conditions, so-called α-testing. Over several weeks, groups of chemists and other professionals conduct the testing, using test cases defined for each person in a test book that must be signed off by individuals on completion of the test. The objective is to test the complete computerized system for functionality, usability, reliability, performance and supportability as stated in the requirement specifications document.

Test cases that will be handed out to the test person are prepared by the quality assurance (QA) department together with the development team. The documents include the scope of the test with a link to the requirement specifications document, the system configuration, background information on the function to be tested, detailed test description and expected results. During and after testing, the test person completes the sheet with his name, actual results and any comments on findings.

The system is not only tested under typical operating conditions but also at the limits under which it will be required to operate – an approach known variously as *worst case* testing, *gray box* testing or testing of *boundary conditions*. Testing boundary conditions is important because most software errors occur around its boundary limits. Combinations of several worst cases are also tested. For example, if a system is specified to acquire data from multiple instruments and the data acquisition rate can be varied, test cases include acquisition from the maximum number of instruments at the highest data rate.

Software testing also includes so-called stress testing. Inputs with inappropriate character types (alphabetic characters instead of numeric ones, for example, or inappropriate character length and character composition) are made, and instrument parameters that lie outside the instrument's operational limits are entered. The expectation is that these inputs will not damage data or disrupt system and software operation and that the system will recover after producing error messages.

The test environment reflects as many system configurations as possible. This includes different equipment that is controlled by the computer, different peripherals, such as printers, CD ROMS, different internal memory (RAM), and different operating systems, for example, Windows 95 and NT.

Test cases reflect typical user applications with manual interactive and automated operation. Automated sequences typically run over 24 h or more, where methods are changed between runs. Data files with different file sizes are generated to make sure that system can handle large files.

The user manual is prepared before the α-test to allow test personnel to verify its accuracy and usefulness. At least one test case requires installation of the software and hardware according to the installation instructions.

Defect tracking system

Documenting software errors is important and problems should not be casually reported for repair by the programmer on an ad hoc basis. Problems found during testing are tracked using the HP internal defect control system (DCS).

Defects are classified by the test person as low, medium, serious and critical. Defect density, discovery rate and defect summaries are recorded for each test cycle and for the entire α-test. Summary sheets include information on version number, test hours, number of defects, total number of defects, defects per hour and the linear fit. An example is shown in Table 1.

The test results are evaluated using the Shooman plot. The discovery rate is plotted versus the total number of defects discovered (Fig. 5). A regression linear fit curve is calculated and plotted together with maximum and minimum fits which by definition have a confidence interval of 5%. From the Shooman reliability model (Fig. 5), the number of remaining defects can be estimated. This information is useful to forecast the number of test cycles that are still necessary and a possible release date.

The number of critical defects after the last test cycle must be zero for the software to pass the release crite-

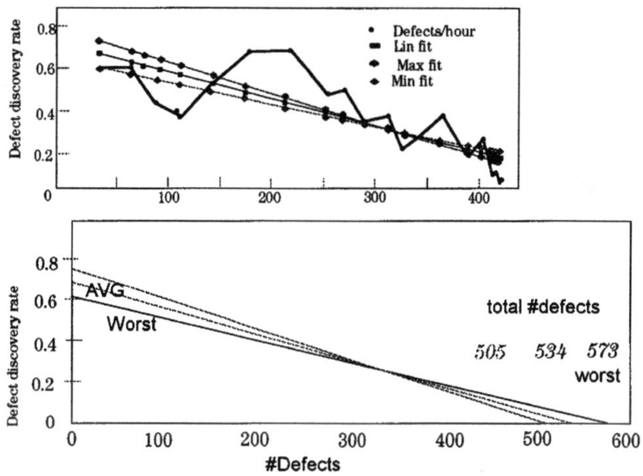

Fig. 5 Example for Schooman Plots and Reliability model with defect discovery rate versus number of defects. Total number of defects is estimated for average (AVG), best and worst case

Table 1 Extract from an alpha-test summary sheet

Version	Test time (hours)	Defects	Cumulative sum of defects	Defect discovery rate (defects/hour)	Line fit
Vxxx	85	80	80	0.94	1.05
	73	69	149	0.86	1.04
	96	79	232	0.82	1.03
↓					
Vyyy	78	3	472	0.04	0.23
	87	3	475	0.03	0.23
	48	1	476	0.02	0.23

The test results are evaluated using the Shooman Plot (10). The discovery rate is plotted versus the total number of defects discovered (Fig. 5). A regression linear fit curve is calculated and plotted together with maximum and minimum fits which by definition have a confidence interval of 5%. From the Shooman reliability model as shown in figure 5 the number of remaining defects can be estimated. This information is useful to forecast the number test cycles that are still necessary and a possible release date.
The number of critical defects after the last test cycle must be zero for the software to pass the release criteria. This and other release criteria are specified in the quality plan.

ria. This and other release criteria are specified in the quality plan.

β-Testing

Once software defects and usability discrepancies reported during α-testing have been corrected, the software may be tested at selected customers' sites (the so-called β-test). The key feature of β-testing is that it is conducted in a customer environment and supervised by a person not involved in the development process. One of the objectives of β-testing is to test the HP product delivery and support channel. A trained HP applications engineer (AE) assists the customer with installation and checks the software installation procedure.

Deliverables for the test phase include:

- Test plans with acceptance criteria and test cases
- Test results
- Validation documents
- Defect density report
- User training material
- System status bulletin (SSB).

Release for production and installation

After the testing is complete and the code is corrected for errors, the software is released for production and distribution. The product is considered ready for release when it has met all the criteria specified in the quality plan and after formal sign-off by product line, quality assurance, and manufacturing management. A prerequisite for this is sufficient training of service engineers who must be able not only to install and operate the software but also to train users and answer users' questions. Availability of user documentation in the form of on-line help and printed reference material is also a release criterion.

The manufacturing department ships the product in accordance with manufacturing guidelines, based on the receipt of valid purchase orders. The product documentation includes a *Declaration of System Validation* with statements from Hewlett-Packard that the software was developed and tested according to the Hewlett-Packard Analytical Software Life Cycle, a process that has been certified for ISO 9001 quality standard and for the information technology quality system (ITQS) compliance and that has been inspected by representatives from both computer validation companies and the pharmaceutical industry.

Summary

Software development and validation activities follow the product life cycle concept where development is divided into different phases. Each phase is completed, reviewed and signed off by management before the next phase starts. The life cycle starts with a proposal and investigation phase where the user requirements and functional specifications (external reference specifications) are set. During the design phase, the design specifications that define internal code structure and formulas are developed. In the implementation phase, the code that meets user requirements and design specifications is written. During all phases, the documents are formally reviewed by peers and correct function of the software is tested by anticipated users in the test phase. Test activities follow a test plan with test cases for each function as specified in the external reference specifications document. Software release for distribution follow criteria which are specified in a quality plan.

References

1. Bedson P, Sargent M (1996) Accred Qual Assur 1:265–274
2. Freeman HM, Leng M, Morrison D, Munden RP 1995 Pharm Tech Europe 40–46
3. Huber L Equipment qualification in practice (to be published)
4. Burgess C (1996) Eur Pharm Rev 52–54
5. Huber L, Welebob L (1997) Accred Qual Assur 2:316–322
6. DIN EN ISO 8402 – Definitions: Quality Management, in English, French and German, August 1995, Beuth, Berlin, Germany
7. Huber L (1996) LC-GC magazine 14:502–508
8. Huber L (1995) Validation of computerized analytical instruments, Interpharm, Buffalo Grove, IL, USA, ISBN 0-935184-75-9 Hewlett-Packard Part Number:5959-3879
9. OECD (1992) Compliance of laboratory suppliers with GLP principles, Series on principles of good laboratory practice and compliance monitoring, No. 5 GLP consensus document, environment monograph No. 49, Paris
10. Shooman M (1973) Operational testing and software reliability estimation during program development, Proc. Int. Symp. Computer Reliability, pp 51–57

Accred Qual Assur (1997) 2:375–380

M. Buzoianu
Hassan Y. Aboul-Enein

Clinical reference materials for the validation of the performance of photometric systems used for clinical analyses

Received: 23 April 1997
Accepted: 7 July 1997

M. Buzoianu
National Institute of Metrology, Sos. Vitan-Bârzesti No.11, 75669 Bucharest, Romania
Tel.: +40 1 634 40 30;
Fax: +40 1 330 15 33

H. Y. Aboul-Enein (✉)
Biological and Medical Research Department (MBC-03), King Faisal Specialist Hospital & Research Centre, P.O.Box 3354, Riyadh 11211, Saudi Arabia
Fax: +966-1-4427858;
e-mail: enein@kfshrc.edu.sa

Abstract There are a wide variety of spectrophotometric devices nowadays used in health services with various qualities of manufacture methods of measurement and metrological characteristics for performing the necessary measurements. Therefore, to meet the accuracy and repeatability requirements needed in medical diagnosis and treatment, the validation of the performance of such systems by clinical chemistry laboratories is essential. However, the validation of a spectrophotometric system for clinical analyses requires several reference materials, according to the end use of the measurement results. This paper discusses some characteristics required of the clinical reference materials needed and used by Romanian Institute of Metrology for validation work. Types of clinical reference materials developed in the national area for this purpose are also presented.

Key words Clinical spectrophotometry · Quality assurance · Validation · Reference materials · Spectrophotometric standards

Introduction

The measurement of components of biological samples faces several problems which influence the quality of the final outcome. Thus, quality control becomes a practice of increasing importance in clinical chemistry.

As is well known, the quality of a measurement result depends on different factors, such as sampling, the measurement method, the instrumentation and the reference materials used for the calibration, etc. This is why it is important that all critical parameters of the instrument and the reference materials should be calibrated and regularly controlled. Experience showed that some manufacturers may not offer sufficient field calibration and testing, and may not be willing to release calibration functions or allow access to raw measurement data. So, as the complexity of instruments increases, it is essential to establish other means to reliably calibrate and control instrumentation. On the other hand, as the instruments used in clinical chemistry laboratories are most diverse and have different degrees of accuracy and repeatability, the validation of the photometric devices used for human blood analyses is very important and currently a subject of considerable interest.

According to the Romanian laws and ordinances issued lately in the field of metrology, all the instruments used in public health are subject to pattern approval, to verification or to compulsory calibration. In this framework, the validation work has a special place.

Generally, validation is a process by which a sample, measurement method, instrument, or piece of data is deemed to be useful for a specific purpose.

The validation of an instrument demands a procedure and a number of standards (reference materials included) with the appropriate characteristics. The definition and use of standards and reference materials available to the clinical chemistry is still a problem in this field. Although there are several clinical reference materials, the problems of using them should not be underestimated (stability, preservation, contamination, use of aqueous solutions, etc.).

A large number of routine clinical chemistry laboratories perform analyses with multichannel analyzers of both the continuous-flow and the direct-sample type. Various types of spectrophotometric reference materials have been recommended to validate the photometric accuracy and linearity, wavelength accuracy or stray light radiation of photometric systems used for clinical analyses. In this respect, much has been done in the national area. The problem we are facing now concerns the required reference materials to be used for the validation of the concentration accuracy of this instrumentation, so widely used in clinical chemistry laboratories.

Aqueous standard solutions containing a mixture of all substances commonly being determined in routine clinical chemistry laboratory are not available. Multichannel systems are therefore being calibrated with commercially supplied calibration sera for which values for each analyte have been assigned by the manufacturers. The same manufacturer also provides control sera for the validation of the system. Wide differences may occur between values obtained on the same analyzer with sera from different manufacturers and even between lots from the same manufacturer (due to the variation in weight or homogeneity of material). In addition, the 'certificates' of the calibration and/or control sera do not often provide information about uncertainties, the method of measurement or the influence parameters. Usually, only the target value or the range within which it lies is indicated for each analyte.

The role of the Standard Reference Materials of the National Institute of Standards and Technology (NIST SRMs) is clearly defined in the validation work. An alternative is to use the certified sera that have been assigned using NIST SRMs. It was of interest to find that the major requirements for concentration reference materials used to validate the photometric systems for clinical analyses have not been well specified in major regulatory guidelines [1]. It seemed appropriate, therefore, in this communication to present some aspects of the Romanian experience on this topic.

The validation of the instrumental performance of photometric systems used for clinical analyses

The quality of the end result fundamentally depends on the sample, the data, the measurement process, the transformation of data to information and, finally, the display and transformation of the information. It is obvious that the role of the instrument in providing the integrity of data is also fundamental to the end result. Written requirements for instrumental performance are not sufficient for assuring the reliability of the result unless they are tested and validated.

The technical specifications identified and described by most of the manufacturers of absorption photometers for medical use include wavelength accuracy, spectral half-width of spectral radiation flux at the detector, photometric accuracy, percentage of wavelength integrated, false radiation, and photometric short-time repeatability. As discussed previously [2], the Instrumental Performance Validation Procedures, issued by serious manufacturers of analytical instruments, indicate the methods and the reference materials required to test and to maintain optimum spectrometer performance in daily routine analysis.

When using automatic continuous flow instrumentation, identifying and describing the performance of the photometric system is rather more difficult. The basic photometric requirements seem clear enough:

1. Given a chemistry which is linear, it is expected that the photometric output would exhibit basic conformance to the Beer-Bouguer Law over the absorbance range 0–2.
2. The sensitivity should be such that the desired or optimum absorbance concentration relationship is achieved and sustained from determination to determination.

In our experience, the various automatic photometric systems yield adequate linear calibration lines through an absorbance of one. However, linearity should not be considered apart from sensitivity, since considerable flexibility exists with regard to the possible ratios of sample to reagent. In this respect, for identification of such performance requirements, this is not a sufficient prerequisite nor a pragmatic way to validate the instrument, since the automatic flow and the specific method and reagents used prevent testing of the the accuracy and repeatability of the photometric system in routine work. Furthermore, the output signal displayed is a concentration value instead of a photometric quantity. As recommended in [3], such evaluation of the instruments for automatic analysis in clinical biochemistry laboratories should take into account the accuracy and repeatability of the concentration measurement using at least five different analytes from appropriate sera reference materials.

For this reason, it is necessary to focus on the meaning of the term 'appropriate' when applied to reference materials and also on the practical procedure in the metrological assurance of clinical concentration measurements in the national area.

Clinical reference materials for the validation of the performance of photometric systems used for clinical analyses

With their values known, according to De Bièvre [4], clinical reference materials are designated:

- to validate a measurement procedure (i.e. to assure that the procedure that includes chemical preparation, instrument calibration, measurement, data acquisition and data treatment, is suitable and performs properly);
- to validate the measurement instrument (i.e. to assure that it provides, properly and reproducibly, stable and accurate information);
- to validate the data acquisition and data treatment of the measurement procedure (i.e. to assure that the associated software leads to the proper data).

This validation process, required prior to measuring any unknown sample, enables the evaluation of uncertainty contributions to the measurement procedure which are present in the measurement of real life samples.

Photometric devices used for clinical analyses are calibrated against calibration sera, usually commercially supplied with the instrument, for which values for each analyte have been assigned by the manufacturers. Wide differences may occur between values obtained on the same device with sera from different manufacturers and even between lots from the same manufacturer (due to the variation in weight or homogeneity of material). On the other hand, great care should be taken in the use of the information concerning the exact method used to assign values to the material.

SRMs supplied by NIST make the verification of commercial calibration sera possible if referee methods of analysis are available for use in conjunction with the SRMs, so the first important issue here is how to achieve traceability of the certified value of the clinical reference materials to the SI unit. Several traceability schemes for reference materials in analytical measurements have been recommended [4, 5]. Closely connected is the problem of the evaluation of the uncertainty of the clinical reference material. In this respect, a major aspect is the significance of this uncertainty in the medical treatment or diagnosis, as it is widely recognized that it is rather difficult to show clearly and unambiguously a limit of accuracy required for a clinical result.

An attempt to evaluate the contribution of the uncertainty of the reference material in the overall uncertainty of the photometric measurement process is illustrated in Fig. 1. Note that the steps followed for evaluating and expressing the measurement uncertainty are in accordance with the metrological approach recently described by Bunzoianu and Aboul-Enein [2].

Figure 1 represents four examples of the evaluation of measurement uncertainty for potassium, calcium, magnesium and glucose using flame photometry, atomic absorption spectrometry and molecular spectrometry (Mg determination with Titan Yellow and glucose determination with glucose oxidase). For the sake of simplicity in Fig. 1, the component of uncertainty due to the sampling is not represented. Thus, the uncertainties of concentration per concentration (u_c/c) due to the photometric devices (1), to the reference materials used for calibration (2), to the calibration (3) and to the volumetric means of measurement (4) are illustrated.

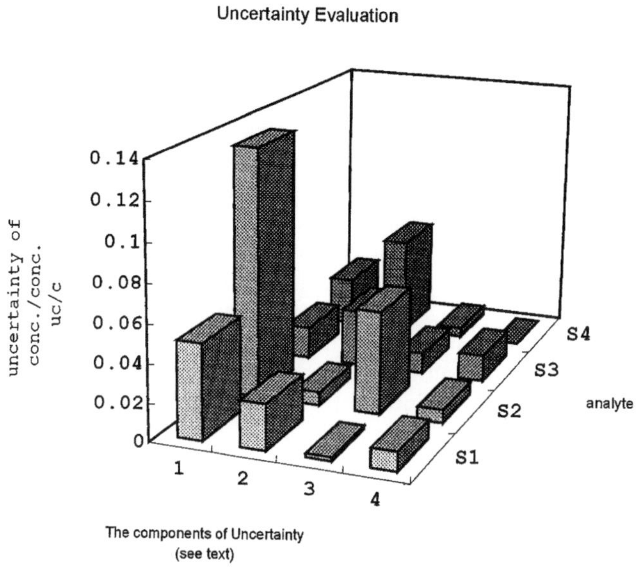

Fig. 1 Measurement uncertainty components for the determination of potassium by flame photometry (*S1*), calcium by atomic absorption spectrometry (*S2*), magnesium by molecular spectrophotometry (*S3*), glucose by molecular spectrophotometry (*S4*)

It should be noted that when we used methods of measurement needing inorganic reference materials for calibration (such as flame photometry or atomic absorption spectrometry) the uncertainty due to the reference materials was considerably lower than that due to the photometric device. On the contrary, when we used a clinical reference material certified for its glucose concentration with a 10% (rel) uncertainty, this uncertainty exceeded twice the uncertainty due to the spectrophotometric device. When we determined Mg by a spectrophotometric method with Titan Yellow, we found that the uncertainty due to the reference material was approximately twice that due the device, as we used a very accurate spectrophotometer.

The clinical reference materials developed for the validation of the performance of photometric systems

The major purposes followed by the Romanian Institute of Metrology in the development of synthetic clinical reference materials were: (1) to be able to handle large quantities of base material in a short time, (2) to prepare samples of correct and adequate size, (3) to prepare samples of the highest quality, (4) to insure, as far as possible, the traceability of the clinical measurements direct to SI units.

Preparation of the clinical reference materials

Four types of reference materials for the validation of the instrumental performances of the photometric devices used for clinical analyses were gravimetrically prepared, under the responsibility of the Romanian National Institute of Metrology, from high-purity reagents as a synthetic material composed of deionized water (electrical conductivity 0,5 μS/cm, sterilised by 0,2-mm filter and UV continuous-flow filter) and suitable inorganic salts containing declared cations, metals and organic salts with choride anions. In general, Certified Reference Materials are expensive consumables. Due to the risk that samples may suffer from instability or degradation or are contaminated, the RMs developed were bottled in 10 ml ampoules made of glass, purified and pre-treated, intended for a limited term (one year maximum). In fact the stability was checked for each type of reference material over 20 months.

The concentrations, in mmol/l, of different components were checked by classical, visible (VIS) spectrophotometry, ion-selective electrometry, atomic absorption spectrometry, continuous flow and flame photometry. The matrix and the molar concentration of the reference materials developed are shown in Table 1.

Table 1 Concentration values: synthetic type reference material sera

Element	Concentration in mmol/dm^3			
	Type 14.01	Type 14.02	Type 14.03	Type 14.04
Na	148.5	126.5	158.4	134.4
K	3.7	6.2	4.2	3.5
Ca	2.3	3.3	3.6	3.0
Cl	170.0	150.0	180	150.0
Mg	0.86	1.8	1.4	1.2
Glucose	5.4	16.5	7.4	6.3
Urea	5.5	19.5	11.4	9.6

Note that the molar concentration values are relevant to decision levels where accurate measurement is crucial for medical decision. The category of synthetic-type reference material sera needs to be certified by means of primary methods. According to international regulations [6], the methods of certification of the clinical reference materials followed several steps regarding the preparation of the material, homogeneity testing, performance of interlaboratory analyses, assignment of the certified value and uncertainty.

Table 2 Concentration values: synthetic type reference material. Grav. M, gravimetric method; Flamp. M, flame photometry; Ion S. M, ion selective method; Cont. F, continuous flow; Spp. M, spectrophotometric method; AAS. M, atomic absorption spectrometry; Vol. M, volumetric method; Gl. ox. M, glucose oxidase method; *o*-Tol. M, ortho-toluidine method

Element		Concentration in mmol/dm^3			
		Type 14.01	Type 14.02	Type 14.03	Type 14.04
Na	Grav. M	148.8 ±0.4	126.8 ±0.2	158.7 ±0.4	134.4 ±0.3
	Flamp. M	148.0 ±4.0	126.0 ±2.0	158.0 ±7.0	135.0 ±3.0
	Ion S. M	149.5 ±5.0	127.0 ±2.0	159.0 ±9.0	136.2 ±3.0
	Cont. F	149.0 ±4.0	127.3 ±3.0	159.9 ±8.0	137.0 ±4.0
K	Grav. M	3.70±0.30	6.20±0.20	4.20±0.20	3.50±0.20
	Flamp. M	3.72±0.20	6.10±0.50	4.30±0.30	3.70±0.20
	Ion S. M	3.75±0.20	6.05±0.40	4.15±0.20	3.65±0.20
	Cont. F	3.85±0.30	6.15±0.50	4.50±0.30	3.55±0.20
Ca	Grav. M	2.30±0.20	3.30±0.10	3.60±0.10	3.00±0.12
	Spp. M	2.40±0.10	3.28±0.10	3.75±0.10	3.10±0.10
	Cont. F	2.60±0.04	3.35±0.06	3.20±0.06	3.09±0.05
	AAS. M	2.26±0.30	3.25±0.30	3.55±0.30	2.90±0.30
	Ion S. M	2.30±0.08	3.40±0.10	3.65±0.10	3.15±0.09
	Vol. M	2.12±0.10	3.20±0.20	3.50±0.20	2.85±0.20
Mg	Grav. M	0.86±0.02	1.80±0.02	1.41±0.01	1.21±0.01
	Spp 1. M	0.84±0.07	1.79±0.08	1.38±0.06	1.19±0.07
	Spp 2. M	0.90±0.07	1.87±0.08	1.42±0.06	1.22±0.07
	Ion S. M	0.86±0.04	1.85±0.04	1.44±0.03	1.27±0.03
	AAS. M	0.88±0.15	1.83±0.15	1.39±0.15	1.25±0.15
Gl	Grav. M	5.40±0.05	16.70±0.20	7.47±0.08	6.36±0.06
	Gl. ox. M	5.43±0.58	14.63±1.00	7.73±0.65	6.60±0.67
	o-Tol. M	5.60±0.34	15.98±0.64	7.85±0.48	6.73±0.43
	Cont. F	5.27±0.30	16.26±0.41	7.22±0.38	6.25±0.32

For each analyte, mean concentrations and their 95% confidence interval were evaluated for each method employed, because differences among the results obtained by the individual methods were significant. The assigned concentration values were established based on the interpretation of data respecting the methods of preparation and measurement (taking into account the contribution of a statistical as well as a systematic nature). Accordingly, the mean values assigned for different methods of measurement for five selected analytes from the synthetic sera reference materials developed are presented in Table 2.

Evaluation and uncertainty

The uncertainties of the assigned values were then evaluated [7] on the basis of the estimation of the Type A and Type B uncertainty components, due to the possible errors originating principally from preparation as well as from the direct measurement (method included), as follows:

$$U_c = k \cdot \sqrt{u_{Ac}^2 + u_{Bc}^2} \quad (1)$$

where U_c is the expanded uncertainty of the concentration *(c)* value, and k is control for the input quantity determined from independent repeated observations. The uncertainty of the above estimate is the Type A standard uncertainty.

$$u_{Ac} = \sqrt{\frac{\sum_{i=1}^{n} \left(c_i - \bar{c} \right)^2}{n-i}} \quad (2)$$

where the c_i is the value of the measured concentration and $\bar{c}$ is the mean of the n measurements performed under repeatable conditions.

For an estimate of an input quantity that has not been obtained from repeated observations, the associated estimated variance or standard uncertainty is evaluated by scientific judgement based on all available information on the possible of its variability. This is the case of Type B standard uncertainty. The pool of information may include previous data, experience with general knowledge of behaviour of relevant materials and instruments, manufacture's specifications, or data provided in calibration and other certificates or uncertainties assigned to references data taken from handbooks.

$$u_{Bc} = \sqrt{\sum_{j=1}^{m} c_{c,xj}^2 \cdot u_{B,xj}^2} \quad (3)$$

where $C^2_{c,xj}$ are the functions describing the concentration estimation depending on x_j influence factors that depend on the method used to certify each element

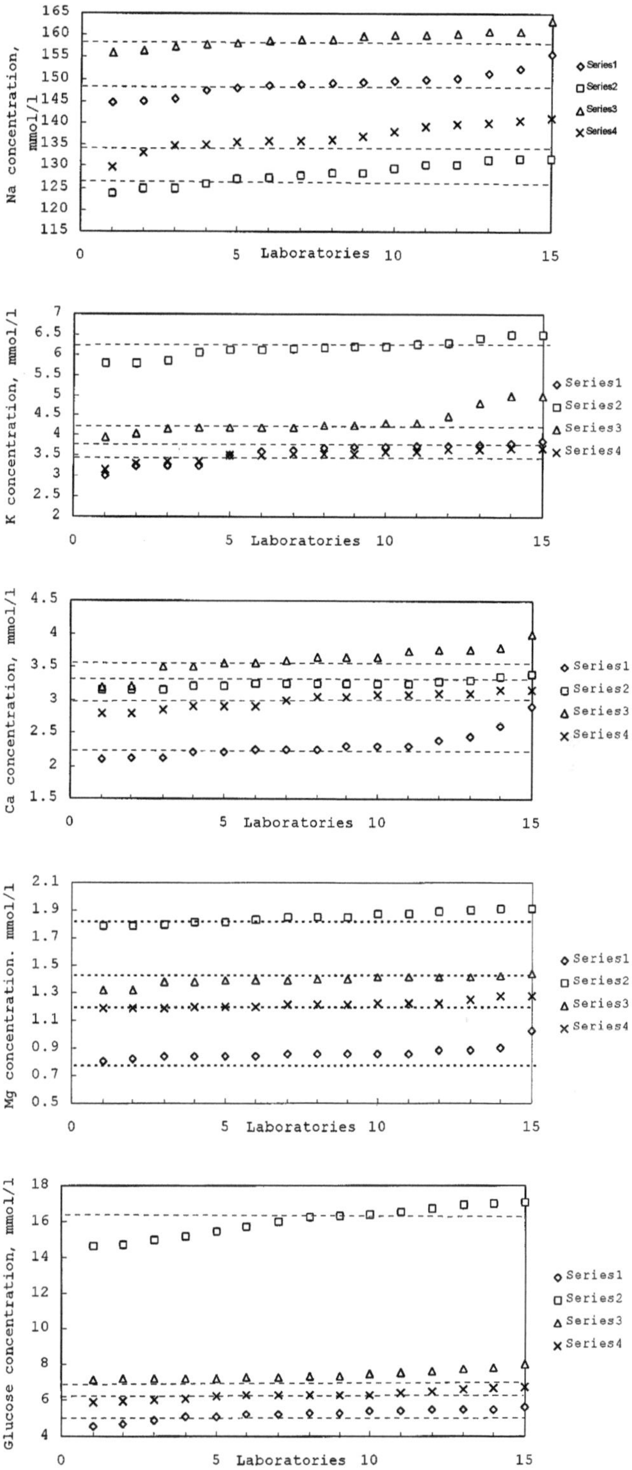

Fig. 2 Analytical data obtained in interlaboratory comparison RM type 14.01–14.04

from the reference material developed (such as mass measurement, volume measurement, spectrophotometric measurement, reference materials used for the calibration, etc.) and $u^2_{B,xj}$ are the uncertainties associated with each of these functions.

Each standard uncertainty for all inputted quantities, evaluated as Type A or Type B procedure, were combined in the correct mathematical manner to evaluate combined standard uncertainty, $u_c^2(y)$, – that characterizes the dispersion of the values that could reasonably be attributed to the considered measurand.

The additional measure of uncertainty that meets the requirement of providing an interval of the above is termed expanded uncertainty and was obtained by multiplying the combined standard uncertainty by a coverage factor k ($k = 2$). Thus, the uncertainties obtained for the assigned values for sodium, potassium, calcium, magnesium and glucose concentration in the reference materials for each methods of measurement used are illustrated in Table 2. Narrower uncertainty limits were obtained for these conventionally certified values in comparison to values of acceptable range stated for commercially available control sera of the same matrix. Furthermore, a comparison was made of the analyses of the synthetic clinical reference materials carried out by 15 selected laboratories (according to the equipment and training in the field of clinical chemistry and the geographic coverage area, respectively) using their respective routine analytical methods and instrumentation. Each laboratory tested the synthetic sera as unknown samples after the calibration of their specific measurement means with control sera supplied from different sources. Whenever possible, the flame photometers, the photometers and the spetrophotometers were previously tested in accordance with the Legal Metrology Norms, issued in accordance with the requirement of the International Organization for Legal Metrology (OIML). The results were graphically compared with the assigned values as shown in Fig. 2. Note that the uncertainty of a value is considerably smaller than the interlaboratory spread. The results showed the following spread: less + 4,9% for Na, less + 19% for K, less + 26,1% for Ca, less + 18,6% for Mg and less - 15,6% for glucose, asymmetrically distributed around the assigned values. The graphs indicate what is the best possible result obtainable on the assigned value and what is obtained in practice under routine conditions. All methods used produced results with roughly the same spread. It may be seen that all the elements tested show some outliers at the high, then on the low concentration side.

No method was found to be superior. The best method used can give wrong results if incorrectly applied, even in the experimental laboratories. Therefore, it is better to impose good quality on a laboratory than a specific method.

The main reasons for the current lack of comparability at working level include an insufficient awareness of uncertainty and source of error, a lack of high quality reference materials and no recognized system for intercomparison of traceable clinical chemistry measurements. In addition, the limit results were obtained in the absence of a reliable uncertainty budget and insufficient QA procedures. In fact, in the national area the introduction of adequate QA procedures is making its first steps.

Conclusion

This paper discusses a number of practical problems arising from the request for and use of clinical reference materials for the validation of the performance of photometric systems used in national clinical chemistry laboratories. It shows that uncertainties in the measurement step of photometric analysis have largely been ignored. Uncertainties associated with this step can and do contribute significantly to overall analytical uncertainty. Thus, for a knowledge of trueness and measurement uncertainty, an adequate certified reference materials system and an attempt at a traceability chain are of the utmost importance, since the quality of clinical chemistry results depends critically on the use of reliable reference materials and properly validated instruments.

References

1. Bronder Th, Rinneberg H (1992) PTB Laborbericht, Proposal for international recommendation
2. Buzoianu M, Aboul-Enein HY (1997) Accred Qual Assur 2:11–17
3. Broughton PM, Buttolph MA, Gowenlock AH, Neill DW, Scentelbery RG (1969) J Clin Path 22:278–284
4. De Bièvre P (1996) Accreditation and quality assurance in analytical chemistry. Springer, Berlin Heidelberg New York, pp 159–193
5. Marschal A (1980) Bull BNM 39:33–39
6. Guide ISO 35 (1989) Certification des materiaux de référence – Principes généraux et statistiques
7. Guide to the expression of uncertainty in measurement (1993) International Organization for Standardization (ISO), Geneva

Accred Qual Assur (1998) 3:6–10

Stephen L.R. Ellison
Alex Williams

Measurement uncertainty and its implications for collaborative study method validation and method performance parameters

Received: 22 September 1997
Accepted: 6 October 1997

S.L.R. Ellison (✉)
Laboratory of the Government Chemist,Queens Road, Teddington, Middlesex, TW11 0LY, UK

A. Williams
19 Hamesmoor Way,Mytchett, Camberley, Surrey, GU16 6JG, UK

Abstract ISO principles of measurement uncertainty estimation are compared with protocols for method development and validation by collaborative trial and concomitant "top-down" estimation of uncertainty. It is shown that there is substantial commonality between the two procedures. In particular, both require a careful consideration and study of the main effects on the result. Most of the information required to evaluate measurement uncertainty is therefore gathered during the method development and validation process. However, the information is not generally published in sufficient detail at present; recommendations are accordingly made for future reporting of the data.

Introduction

One of the fundamental principles of valid, cost-effective analytical measurement is that methodology should demonstrably fit its intended purpose [1]. Technical fitness for purpose is usually interpreted in terms of the required "accuracy". To provide reasonable assurance of fitness for purpose, therefore, the analyst needs to demonstrate that the chosen method can be correctly implemented and, before reporting the result, needs to be in a position to evaluate its uncertainty against the confidence required.

Principles for evaluating and reporting measurement uncertainty are set out in the 'Guide to the expression of uncertainty in measurement' (GUM) published by ISO [2]. EURACHEM has also produced a document "Quantification of uncertainty in analytical measurement" [3], which applies the principles in this ISO Guide to analytical measurements. A summary has been published [4]. In implementing these principles, however, it is important to consider whether existing practice in analytical chemistry, based on collaborative trial [5–7], provides the information required. In this paper, we compare existing method validation guidelines with published principles of measurement uncertainty estimation, and consider the extent to which method development and validation studies can provide the data required for uncertainty estimation according to GUM principles.

Measurement uncertainty

There will always be an uncertainty about the correctness of a stated result. Even when all the known or suspected components of error have been evaluated and the appropriate corrections applied, there will be uncertainty on these corrections and there will be an uncertainty arising from random variations in end results.

The formal definition of "Uncertainty of Measurement" given by the GUM is "A parameter, associated with the result of a measurement, that characterises the dispersion of the values that could reasonably be attributed to the measurand. Note (1): The parameter may be, for example, a standard deviation (or a given multiple of it) or the half width of an interval having a stated level of confidence."

For most purposes in analytical chemistry, the "measurand" is the concentration of a particular species. Thus, the uncertainty gives a quantitative indication of the range of the values that could reasonably be attributed to the concentration of the analyte and enables a judgement to be made as to whether the result is fit for its intended purpose.

Uncertainty estimation according to GUM principles is based on the identification and quantification of the effects of influence parameters, and requires an understanding of the measurement process, the factors influencing the result and the uncertainties associated with those factors. These factors include corrections for duly quantified bias. This understanding is developed through experimental and theoretical investigation, while the quantitative estimates of relevant uncertainties are established either by observation or prior information (see below).

Method validation

For most regulatory applications, the method chosen will have been subjected to preliminary method development studies and a collaborative study, both carried out according to standard protocols. This process, and subsequent acceptance, forms the 'validation' of the method. For example, the AOAC/IUPAC protocol [5, 6] provides guidelines for both method development and collaborative study. Typically, method development forms an iterative process of performance evaluation and refinement, using increasingly powerful tests as development progresses, and culminating in collaborative study. On the basis of the results of these studies, standard methods are accepted and put into use by appropriate review or standardisation bodies. Since the studies undertaken form a substantial investigation of the performance of the method with respect to trueness, precision and sensitivity to small changes and influence effects, it is reasonable to expect some commonality with the process of uncertainty estimation.

Comparison of measurement uncertainty and method validation procedures

The evaluation of uncertainty requires a detailed examination of the measurement procedure. The steps involved are shown in Fig. 1. This procedure involves very similar steps to those recommended in the AOAC/IUPAC protocol [5, 6] for method development and validation, shown in Fig. 2. In both cases the same processes are involved: step 1 details the measurement procedure, step 2 identifies the critical parameters that influence the result, step 3 determines, either by experiment or by calculation, the effect of changes in each of these parameters on the final result, and step 4 their combined effect.

The AOAC/IUPAC protocol recommends that steps 2,3 and 4 be carried out within a single laboratory, to optimise the method, before starting the collaborative trial. Tables 1 and 2 give a comparison of this part of the protocol [6] with an extract from corresponding parts of the EURACHEM Guide [3]. The two procedures are very similar. Section 1.3.2 of the method vali-

Fig. 1

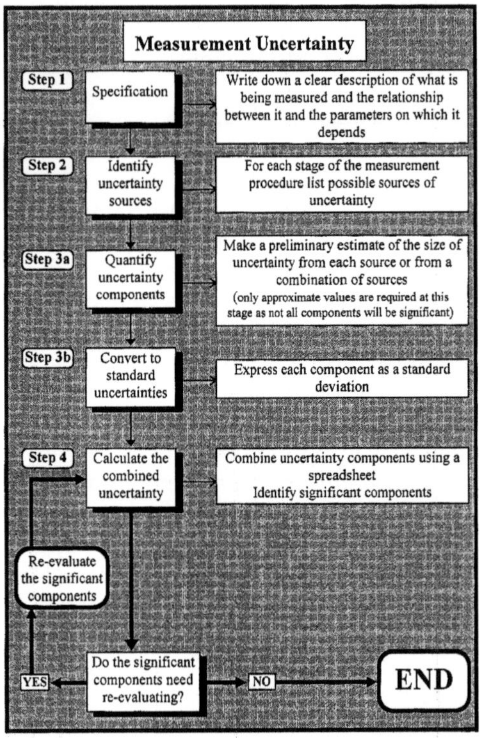

Fig. 2

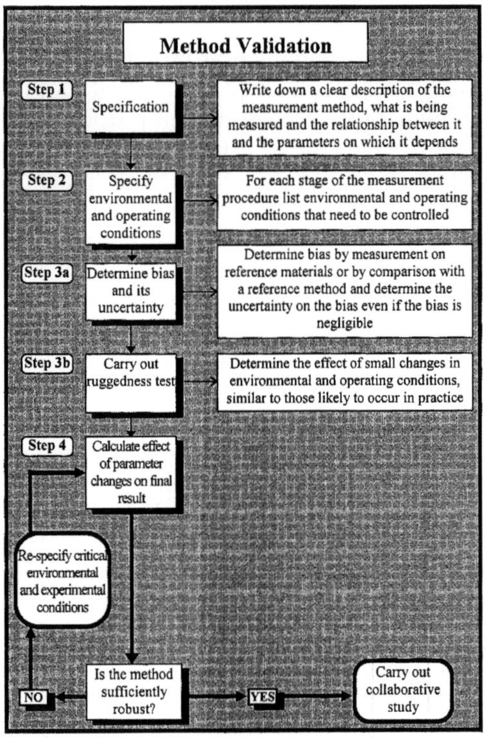

Table 1 Method development and uncertainty estimation

Method validation[1]	Uncertainty estimation
1.3.2 Alternative approaches to optimisation (a) Conduct formal ruggedness testing for identification and control of critical variables. (b) Use Deming simplex optimisation to identify critical steps. (c) Conduct trials by changing one variable at a time.	Having identified the possible sources, the next step is to make an approximate assessment of size of the contribution from each source, expressed as a standard deviation. Each of these separate contributions is called an uncertainty component. Some of these components can be estimated from a series of repeated observations, by calculating the familiar statistically estimated standard deviation, or by means of subsidiary experiments which are carried out to assess the size of the component. For example, the effect of temperature can be investigated by making measurements at different temperatures. This experimental determination is referred to in the ISO Guide as "Type A evaluation".

[1] Reprinted from The Journal of AOAC INTERNATIONAL (1989) 72(4):694–704. Copyright 1989, by AOAC INTERNATIONAL, Inc.

dation protocol is concerned with the identification of the critical parameters and the quantification of the effect on the final result of variations in these parameters; the experimental procedures (a) and (c) suggested are closely similar to experimental methodology for evaluating the uncertainty. Though the AOAC/IUPAC approach aims initially to test for significance of change of result within specified ranges of input parameters, this should normally be followed by closer study of the actual rate of change in order to decide how closely a parameter need be controlled. The rate of change is exactly what is required to estimate the relevant uncertainty contribution by GUM principles. The remainder of the sections in the extract from the protocol give guidance on the factors that need to be considered; these correspond very closely to the sources of uncertainty identified in the EURACHEM Guide. The data from method development studies required by existing method validation protocols should therefore provide much of the information required to evaluate the uncertainty from consideration of the main factors influencing the result.

The possibility of relying on the results of a collaborative study to quantify the uncertainty has been considered [8], following from a general model of uncertainties arising from contributions associated with method bias, individual laboratory bias, and within- and between-batch variations. Collaborative trial is expected to randomise most of these contributions, with the exception of method bias. The latter would be addressed via combination of the uncertainties associated with a reference material or materials to which results are traceable with the statistical uncertainty associated with any estimation of bias using a finite number of observations. Note that the necessary investigation and reporting of bias and associated statistical uncertainty (i.e. excluding reference material uncertainty), are now recommended in existing collaborative study standards [7]. Where the method bias and its uncertainty are small, the overall uncertainty estimate is expected to be represented by the reproducibility standard deviation. The approach has been referred to as a "top-down" view. The authors concluded that such an approach would be feasible given certain conditions, but noted that demonstrating that the estimate was valid for a particular laboratory required appropriate internal quality control and assurance. Clearly, the controls required would relate particularly to the principal factors affecting the result. In terms of ISO principles, this requirement corresponds to control of the main contributions to uncertainty; in method development and validation terms, the requirement is that factors found to be significant in robustness testing are controlled within limits set, while factors not found individually significant remain within tolerable ranges. In either case, where the control limits on the main contributing factors, together with their influence on the result, are known to an individual laboratory, the laboratory can both check that its performance is represented by that observed in the collaborative trial and straightforwardly provide an estimate of uncertainty following ISO principles.

The step-by-step approach recommended in the ISO Guide and the "top down" approach have been seen as alternative and substantially different ways of evaluating uncertainty, but the comparison between method development protocols and ISO approach above shows that they are more similar than appears at first sight. In particular, both require a careful consideration and study of the main effects on the result to obtain robust results accounting properly for each contribution to overall uncertainty. However, the top down approach relies on that study being carried out during method development; to make use of the data in ISO GUM estimations, the detailed data from the study must be available.

Availability of validation data

Unfortunately, the necessary data are seldom readily available to users of analytical methods. The results of the ruggedness studies and the within-laboratory op-

Table 2 Method performance and measurement uncertainty estimation. Note that the text is paraphrased for brevity and the numbers in parentheses refer to corresponding items in the EURACHEM guide (column 2)

Method validation protocol[1]	EURACHEM guide
1.4 Develop within-laboratory attributes of the optimised method (Some items can be omitted; others can be combined.) 1.41 Determine [instrument] calibration function ... to determine useful measurement range of method. (8, 9) 1.4.2 Determine analytical function (response vs concentration in matrix ...). (9) 1.4.3 Test for interference (specificity): (a) Test effects of impurities ... and other components expected ... (5) (b) Test non-specific effects of matrices. (3) (c) Test effects of transformation products ... (3) 1.4.4 Conduct bias (systematic error) testing by measuring recoveries ... (Not necessary when method itself defines the property or component.) (3, 10, 11) 1.4.5 Develop performance specifications ... and suitability tests ... to ensure satisfactory performance of critical steps ... (8) 1.4.6 Conduct precision testing ... [including] ... both between-run (between-batch) and within-run (within-batch) variability. (4, 6, 7, 8, 12) 1.4.7 Delineate the range of applicability to the matrices or commodities of interest. (1) 1.4.8 Compare the results of the application of the method with existing tested methods intended for the same purposes, if other methods are available. 1.4.9 If any of the preliminary estimates of the relevant performance of these characteristics are unacceptable, revise the method to improve them, and retest as necessary 1.4.10 Have method tried by analyst not involved in its development. Revise method to handle questions raised and problems encountered.	The evaluation of uncertainty requires a detailed examination of the measurement procedure. The first step is to identify possible sources of uncertainty. Typical sources are: 1. Incomplete definition of the measurand (for example, failing to specify the exact form of the analyte being determined). 2. Sampling – the sample measured may not represent the defined measurand. 3. Incomplete extraction and/or pre-concentration of the measurand, contamination of the measurement sample, interferences and matrix effects. 4. Inadequate knowledge of the effects of environmental conditions on the measurement procedure or imperfect measurement of environmental conditions. 5. Cross-contamination or contamination of reagents or blanks. 6. Personal bias in reading analogue instruments. 7. Uncertainty of weights and volumentric equipment. 8. Instrument resolution or discrimination threshold. 9. Values assigned to measurement standards and reference materials. 10. Values of constants and other parameters obtained from external sources and used in the data reduction algorithm. 11. Approximations and assumptions incorporated in the measurement method and procedure. 12. Variations in repeated observations of the measurand under apparently identical conditions.

[1] Reprint from The Journal of AOAC INTERNATIONAL (1989) 72(4):694–704. Copyright 1989, by AOAC INTERNATIONAL, Inc.

timisation of the method are, perhaps owing to their strong association with the development process rather than end use of the method, rarely published in sufficient detail for them to be utilised in the evaluation of uncertainty. Further, the range of critical parameter values actually used by participants is not available, leading to the possibility that the effect of permitted variations in materials and the critical parameters will not be fully reflected in the reproducibility data. Finally, bias information collected prior to collaborative study has rarely been reported in detail (though overall bias investigation is now included in ISO 5725 [7]), and the full uncertainty on the bias is very rarely evaluated; it is often overlooked that, even when investigation of bias indicates that the bias is not significant, there will be an uncertainty associated with taking the bias to be zero [9], and it remains important to report the uncertainty associated with the reference material value.

Recommendations

The results of the ruggedness testing and bias evaluation should be published in full. This report should identify the critical parameters, including the materials within the scope of the method, and detail the effect of variations in these on the final result. It should also include the values and relevant uncertainties associated with bias estimations, including both statistical and reference material uncertainties. Since it is a requirement of the validation procedure that this information should be available before carrying out the collaborative study, publishing it would add little to the cost of validating the method and would provide valuable information for future users of the method.

In addition, the actual ranges of the critical parameters utilised in the trial should be collated and included in the report so that it is possible to determine their effect on the reproducibility. These parameters will have been recorded by the participating laboratories, who normally provide reports to trial co-ordinators; it should therefore be possible to include them in the final report.

Of course there will frequently be additional sources of uncertainty that have to be examined by individual laboratories, but providing this information from the validation study would considerably reduce the work involved.

Acknowledgements The preparation of this paper was supported under contract with the Department of Trade and Industry as part of the National Measurement System Valid Analytical Measurement (VAM) Programme [10].

References

1. Sargent M (1995) Anal Proc 32:201–202
2. ISO (1993) Guide to the expression of uncertainty in measurement. ISO, Geneva, Switzerland, ISBN 92-67-10188-9
3. EURACHEM (1995) Quantifying uncertainty in analytical measurement. Laboratory of the Government Chemist, London. ISBN 0-948926-08-2
4. Williams A (1991) Accred Qual Assur 1:14-17
5. Horwitz W (1995) Pure Appl Chem 67:331–343
6. AOAC recommendation (1989) J Assoc Off Anal Chem 72:694–704
7. ISO (1994) ISO 5725:1994 Precision of test methods. ISO, Geneva
8. Analytical Methods Committee (1995) Analyst 120:2303–2308
9. Ellison SLR, Williams A (1996) In: Parkany M (ed) The use of recovery factors in trace analysis. Royal Society of Chemistry, London
10. Fleming J (1995) Anal Proc 32:31–32

Accred Qual Assur (1998) 3:2–5

Ludwig Huber

Qualification and validation of software and computer systems in laboratories

Part 2: Qualification of vendors

Received: 8 August 1997
Accepted: 12 September 1997

L. Huber (✉)
Hewlett-Packard Waldbronn,
P.O. Box 1280,
D-76337 Waldbronn, Germany
Tel.: +49-7243-602 209
Fax: +49-7802-981948
e-mail: Ludwig_Huber@hp.com

Abstract When software and computer systems are purchased from vendors, the user is still responsible for the overall validation. Because the development validation can only be done by the developers, the user can delegate this part to the vendor. The user's firm should have a vendor qualification program in place to check for this. The type of qualification depends very much on the type and complexity of software and can go from documented evidence of ISO 9001 or equivalent certification for off-the-shelf products to direct audit for software that has been developed on a contract basis. Using a variety of practical examples, the article will help to find the optimal qualification procedure.

Key words Validation · Qualification · Vendors · Computers · Analytical laboratories

Introduction

Software and computer systems should validated and qualified during all phases of their life cycle. The validation and qualification tasks during development have been discussed in an earlier article of this series [1]. This article gives recommendations on how to select and qualify a vendor when software or complete systems are purchased.

When software and computerized systems are purchased from a vendor a frequently asked question is who is responsible for the validation of such a system: the vendor or the user? The Organization for Economic Cooperation and Development (OECD) states clearly in consensus paper No. 5: "It is the responsibility of the user to ensure that the software program has been validated" [2]. This is also the practice of the US FDA (United States Food and Drug Administration) as specified by Tetzlaff, a former FDA investigator: "The responsibility for demonstrating that systems have been validated lies with the user" [3]. However, it is obvious that product quality cannot be achieved by testing in a user's laboratory. This must be incorporated during design and development, which can only be done by the vendor. Therefore the OECD makes a further statement in consensus paper No. 5: "It is acceptable for formal validation to be carried out by the supplier on behalf of the user". Furman et al. [4] of the US FDA also make it very clear: "All equipment manufacturers should have validated their software before releasing it".

A more recent OECD consensus paper [5] requires the development of software in a quality system environment: "There should be adequate documentation that each system was developed in a controlled manner and preferably to recognized quality and technical standards (e.g. ISO 9001)" Similarly the European Union requires in Annex 11 of the EC guide to good manufacturing practice (GMP) for medicinal products [6]: "The user shall ensure that software has been produced in accordance with a system of quality assurance".

The objective of vendor qualification is to get assurance that the vendor's products development and manufacturing practices meet the requirements of the user's firm. For software development this usually means that the software is developed and validated following documented procedures. The requirements usually vary between user firms and, within companies, between different departments. An example of a development and validation procedure as typically required by the pharmaceutical manufacturing has been published by Huber [1] in an earlier article of this series.

The dilemma of the user is that she or he should ensure that the software has been validated during development even if she/he typically has no insight into the vendor's practices and most of the time does not have the technical understanding of how software should be validated during development. So the question is: How can an analytical chemist, a good laboratory practice (GLP) study director, or a laboratory manager decide whether the software she or he is using has been validated during development? Ideally, there should be a software vendor qualification scheme that certifies software vendors to be in compliance with all regulations. Unfortunately, even though some software certification systems exist, e.g., Information Technology Quality System (ITQS) [7] and TickIt [8], none of them is accepted by regulatory agencies as being always sufficient for vendor qualification.

This article should help to get around the dilemma and to establish an efficient vendor qualification program for different software categories with minimum efforts. The computer systems will be classified as standard systems and software or systems that have been developed on a contract basis specifically for the user. Special attention is paid to the requirements of users in regulated pharmaceutical industries.

Standard systems

Vendor qualification is relatively easy if the software or computer system is sold in the same configuration to multiple users. Typical examples are computer systems for analytical instrument control, data acquisition and data evaluation. In this case there is much information available on the quality of the product and on the vendor's behavior in case of problems. This information may be available within the user's firm based on experience of previous versions within the same department or the newly purchased version in other departments. Information can also be gleaned from colleages in other companies. If such information is available, only a little additional information need be collected from the vendor. This can include questions like

1. Has the software been developed and manufactured in accordance with a system of quality assurance? Usually, certification for ISO9001 provides such evidence.
2. Can the vendor provide certified evidence that the software has been validated during development? Certification for ITQS [7] or IckIT [8] provides such evidence. Both systems certify compliance with ISO Guide 9000-3 [9]. The TickIT scheme was jointly developed by the United Kingdom Department of Trade and Industry (DTI) and the British Computer Society. ITQS is an international mutual-recognition scheme for ISO 9000 registration in the software and hardware sections of Information Technology and Telecommunications sector. A major objective of ITQS is to expand the international agreement on the registration of IT companies and to provide a global scheme for the recognition of certification in this area.
3. Can the vendor guarantee access to software development documents? This question is particularly important for those working in the regulated environments. Shipment of a declaration of system validation with each software or computer system provides such evidence. This declares that each software product and their actual revisions have been validated and that additional information on product development can be made available.

In addition, regulatory agencies need to get assurance that the software source code can be made available in case it is required. Therefore, for standard software used in GLP and in GMP environments, vendors should make a statement that the source code can be made accessible at the vendor's site. A statements on this should be included in the vendor's declaration of system validation.

Instrument vendors generally react positively to this. For example, most vendors develop and validate analytical products following documented product life cycles. Products are shipped with a "Declaration of System Validation" or similar documents that certify that the specific product was developed and validated following the product life cycle process. Most vendors are also certified for ISO 9001 and some also for ITQS or TickIt. Some vendors also make further information on development and testing available to the user on special request, and some guarantee accessibility of the source code to regulatory agencies.

If a vendor is not able or willing to provide documented evidence of validation, the user should consider selecting another vendor: "Companies should consider alternative vendors when they encounter suppliers who are unable or unwilling to share test data or evidence to support system performance" [3]. This recommendation is easy to follow if there are a number of competitors for the same or similar products. If this is not the case, for example when special software for an emerging technique is required, a user may purchase the software anyway, but should evaluate the vendor using the criteria usually applied for vendor qualification for non-standard software.

Users of larger computer systems in pharmaceutical development sometimes feel that they should do more

Table 1 Checklist for supplier assessments (from [10])

Audit Item
Company information
Company history: how long has the company been in business?
Financial status (obtain copy of annual report)?
Is the company currently in the process of negotiation for sale?
Size (number of employees?)
What percentage of sales is invested in research and development of new products?
Does the vendor have an established customer base in the user firm's market place?
Are there company policies on quality, security etc.?
Is there a Quality Management system?
Is the vendor compliant to ISO 9001? (obtain copies of certificates)
Is the vendor compliant to ISO Guide 9000-3 (obtain copies of ITQS or TickIT)
Has the company been audited by other companies?
Organization
Is there a formal quality assurance department (ask for an organization chart)?
Software development
Does the vendor follow engineering standards?
Is there a software quality assurance program?
Is there a structured methodology (e.g. life cycle approach)
Are there life cycle checkpoint forms (checklists)?
Is prototyping done before and during development?
Is all development done at the vendor's site?
If not, are third parties certified or regularly audited by the vendor (e.g. ISO 9001?
Testing
Who develops test plans?
Are requirement specifications reviews and design inspections done periodically?
How is functional testing performed?
Who is testing (outside the development department)?
Are there test protocols?
Is there a test traceability matrix to ensure that all user requirements and product functions are tested?
Are there procedures for recording, storing and auditing test results?
Who approves test plans/protocols?
Support/training
How many support personnel does the company have?
Does the vendor have formalized training programs in installation, operation and maintenance of systems?
Which support systems are in place (phone, direct)?
Where is the nearest office for support?
Is a service contract available and what does it cover (installation, startup, performance verification, maintenance, training?)
Does the company provide consulting and validation services?
Do support people speak local language?
What is the average response time?
Is the service organization compliant to an international quality standard (for example, ISO 9002 or ISO 9003)?
How long are previous versions supported and at what level?
How long is support and supply of parts guaranteed?
Is training available on how to use the system? Location, frequency?
Are training materials available (description, media)?
Failure reports/enhancement requests
Is there a formal problem-reporting and feedback system in place?
How are defects and enhancement requests handled?
How are customers informed on failure handling?
Are quality records and statistical failure evaluations in existence?
Change control
Who initiates changes?
Who authorizes changes?
Are there procedures for change control?
Do they include impact analysis, test plans?
Is there a formal revision control procedure?
Will all updates get new version numbers?
Are there procedures for user documentation updates?
How are customers informed on changes?
People qualification
Do people have knowledge of regulatory compliance and programming science?
Is there documentation on education, experience, training?
The product/project
When did development of the software first begin?
When was the initial version of the software first released?
How many systems are installed?
How often are software releases typically issued?
How many employees are working on the project?
Are there functional specifications?
Are there samples of reports?
Which vintage of data files can be processed with today's software?
User documentation
Are there procedures and standards for the development and maintenance of user documentation?
What documentation is supplied?
For how long is the documentation supplied?
For how long is the user documentation archived?
Archiving of software and documentation
What is archived, for how long (software, revisions, source code, documentation)?
Where is the source code archived?
Can the source code be made accessible to regulatory agencies?
Is a periodic check of data integrity ensured?
Security
Is the developer's area secure?
What type of security is provided to prevent unauthorized changes?
Are there written procedures specifying who has access to software development and control?
Customer training
Does a training manual exist?
Do they provide tools for training e.g., computer-based or video training?
Do they offer operator training courses (frequency, language)?
Is there documented evidence for the trainer's qualifications?

detailed qualification than that recommended in previous paragraphs, even for standard systems, and give consideration to a direct audit at the vendor's site, especially when specific software and computer systems are widely used throughout the company. Direct audits are expensive and time-consuming for both the user's and the vendor's firm. Therefore both parties should do their utmost to prevent such audits. The vendor can help by providing the user with appropriate detailed written information on the software quality system. The user can develop checklists usually used during direct audits and send it to the vendor. An example is shown in Table 1. The checklists should be returned within about 3–4 weeks. In our experience, direct audits are not necessary for standard systems if the answers to these questions are positive, and we are not aware of any situation where a user's firm failed a regulatory inspection after vendors of standard software had been qualified as described in this article.

Non-standard software and computer systems

If standard software has been modified or new software has been developed on special request by a user, the vendor should be qualified through audits. This can be done through a detailed written questionnaire similar to the questions in Table 1 for smaller projects or through a detailed site audit (second party audit) . Frequently the quality requirements are agreed upon before the programming is started as part of the purchase agreement. Compliance with the agreement is checked during and after development in direct audits. Questions covered during the audit are similar to those in Table 1. Here it is important to check selected examples of the documentation to make sure that this, too, is satisfactory. Details on how to conduct a software supplier audit have been published by Segalstad [11].

Summary

Software and computer systems must be validated before and during routine use. Also, for purchased software, the user is responsible for the entire validation but the vendor can do part of the validation on behalf of the user. The user should have a qualification process in place to assure that the vendor did develop and validate the software according to quality assurance standards. For standard off-the-shelf systems, certification to a general quality standard, e.g. ISO 9001, provides enough evidence of validation and, for software, certification using a system such as ITQS or TickIT together with the vendor's declaration of software validation with the assurance of accessibility to validation documents and the source code for regulated industry is sufficient. When special software is developed on behalf of the user, quality assurance measures with details of development validation can be part of the purchase agreement. In this case, the user may check compliance with the agreement by means of direct audits.

References

1. Huber L (1997) Accred Qual Assur 2:360–366
2. OECD (1992) Compliance of laboratory suppliers with GLP principles. (Series on principles of good laboratory practice and compliance monitoring, No. 5, GLP consensus document, environment monograph No. 49) Paris
3. Tetzlaff RF (1992) Pharm Tech:71–82
4. Furman WB, Layloff TP, Tetzlaff RF (1994) J AOAC Intern 77:1314–1318
5. OECD (1995) The application of the principles of GLP to computerized systems. (Series on principles of good laboratory practice and compliance monitoring, GLP consensus document No. 10, environment monograph No. 116) Paris
6. EC guide to good manufacturing practice for medicinal products (1992) In: The rules governing medicinal products in the European community, vol 4. Office for Official Publications for the European Communities, Luxembourg
7. European Information Technology Quality System Auditor Guide, January 1992, document 92/0001 (ITAGC) and SQA-68 (CTS-2). Available from ITQS Secretariat, Brussels, Belgium
8. The TickIT guide to software quality management system construction and certification using EN 29001, Issue 2, February 1992. Available from the DISC TickIT Office London
9. ISO 9000–3:1991(E), Quality management and quality assurance standards, Part 3: Guidelines for the application of ISO 9001 to development, supply and maintenance. International Organization for Standardization, Geneve, Switzerland
10. Huber L (1995) Validation of computerized analytical instruments. Interpharm, Buffalo Grove, IL, USA, Hewlett-Packard Part Number: 5959-3879
11. Segalstad S (1996) Eur Pharm Rev: 37–44

Accred Qual Assur (1998) 3:140–144

Ludwig Huber

Qualification and validation of software and computer systems in laboratories

Part 3. Installation and operational qualification

Received: 31 October 1997
Accepted: 25 November 1997

L. Huber (✉)
Hewlett-Packard Waldbronn,
PO Box 1280,
D-76337 Waldbronn, Germany
Fax: +49-7802-981948
Tel.: +49-7243-602 209
e-mail: Ludwig_Huber@hp.com

Abstract Installation and operational qualification are important steps in the overall validation and qualification process for software and computer systems. This article guides users of such systems step by step through the installation and operational qualification procedures. It provides guidelines on what should be tested and documented during installation prior to routine use. The author also presents procedures for the qualification of software using chromatographic data systems and a network server for central archiving as examples.

Key words Validation · Qualification · Computers · Software · Analytical · Laboratories

Introduction

Qualification and validation activities to assess the performance of software and computer systems cover the whole life-cycle of the products. The first article of this series [1] gave an overview of the validation process during development; the second [2] discussed qualification of a vendor when the software or computer system is purchased. This article will discuss the validation steps required during installation and tests required prior to operation – processes called installation qualification and operational qualification (OQ) or acceptance testing respectively. Operational qualification of a computer system is the most difficult qualification task. The reason is that still today there are no guidelines available on what and how much testing should be done. The basic question is: how much testing is enough? Too much testing can become quite expensive, and insufficient testing can be a problem during an audit. For example, we have seen test protocols of 200 and more pages that users of a commercial computerized chromatographic systems have developed over several weeks. Each software function, such as switching the integrator on and off, has been verified as part of an OQ procedure, which is not necessary if such tests have been done to a large extent at the vendor's site.

This article gives practical validation steps for two types of computer systems: integrated computerized analytical systems and a network for file printing and archiving.

Pre-installation

Before a computerized analytical system arrives at the user's laboratory, serious thought must be given to its location, environmental and space requirements. A comprehensive understanding of the requirements of the new computer system must be obtained from the vendor well in advance: required bench or floor space and environmental conditions such as humidity and temperature. Bench space should also include some space left and right of the equipment required for proper operation. Care should be taken that all the environmental conditions and electrical grounding are within the limits specified by the vendor and that the correct cables are used. Environmental extremes of temperatures, humidity, dust, power feed line voltage fluctuations, and electromagnetic interference should be avoided. If environmental conditions could influence the validity of test results, the laboratory should have facilities to monitor and record these conditions, either continuously or at regular intervals.

Table 1 Steps before installation

- Obtain manufacturer's recommendations for installation site requirements.
- Check the site for the fulfillment of the manufacturer's recommendations (space, utilities such as electricity, and environmental conditions such as humidity and temperature).
- Allow sufficient shelf space for SOPs, operating manuals and software

Installation

When the computer system arrives, the shipment should be checked by the user for completeness. It should be confirmed that the equipment ordered is what was in fact received. Besides the equipment hardware, other items should be checked, for example correct cables, other accessories and documentation. A visual inspection of the entire hardware should follow to detect any physical damage. For more complex instrumentation, for example, when multiple computers are connected to a network, wiring diagrams should be produced, if not supplied by the vendor. Distance between the computers and peripherals such as printers and analytical equipment must be within manufacturer's specifications. For example, long low-voltage electrical lines from analytical equipment to computers are vulnerable to electromagnetic interference. This may result in inaccurate input data to the computer. In addition, electrical lines should be shielded if motors or fluorescent light sources are nearby. At the end of the installation of the hardware an electrical test of all computer modules and systems should follow.

Table 2 Steps during installation

- Compare computer system as received with purchase order (including software, accessories, spare parts).
- Check documentation for completeness (operating manuals, maintenance instructions, standard operating procedures for testing, safety and validation certificates).
- Check equipment for any damage.
- Install computer hardware and peripherals. Make sure that distances are within the manufacturer's specifications.
- Swith on the instruments and ensure that all modules power up and perform an electronic self-test.
- Install software on computer following the manufacturer's recommendations.
- Verify correct software installation.
- Make back-up copy of software and installation verification files.
- Configure peripherals, e.g. printers and equipment modules.
- Identify and make a list with description of all hardware, include drawings where appropriate.
- Make a list with description of all operating and applications software installed on the computer.
- List equipment manuals and SOPs.
- Prepare an installation report.

Computer hardware should be well documented with model number and serial and revision numbers, and software should be documented with model and revision numbers. For larger laboratories with many computer systems this should preferably be documented via a computer-based database.

Documentation should include items such as size of the hard disk, internal memory (RAM), installed type and version of operating software, standard application software and user-contributed software, e.g., Macro programs. This information is important because all items can influence the overall performance of a computer system. The information should be readily available in case a problem occurs with the computer system. Table 3 includes items for proper documentation of computer hardware and software.

Table 3 Form for computer system identification

Computer hardware
- Manufacturer, model
- Serial number
- Processor
- International memory (RAM)
- Graphics adapter
- Hard disk (type, partitions, memory sizes)
- Installed drives
- Pointing device (e.g., mouse)
- Space requirement

Monitor
- Manufacturer, model
- Serial number

Printer
- Manufacturer, model
- Serial number
- Space requirement

Instrument interface card
- Type, select code, slot number

Connected equipement hardware
- Hardware module 1
- Interface card setting

Modem
- Type, speed

Network connection
- Card type
- Network address

Operating software
- Operating system (version)
- User interface (version)

Application software 1
- Description
- Manufacturer/vendor
- Product number (version)
- Required disk space

Application software 2
- Description
- Manufacturer/vendor
- Product number (version)
- Required disk space

When complex software is installed on a computer, the correctness and completeness of the installed program and data files should be verified. The problem is that wrong or incomplete software installation may be identified as such not at installation or during initial testing but during routine use, when that particular program is used. Vendors can assist installation verification by supplying installation reference files and automated verification procedures. In this case, the integrity of each file is verified by comparing the cross-redundancy-check (CRC) of the installed file with the checksum of the original file recorded on the installation master. Modified or corrupt files have different checksums and are thus detected by the verification program. Verification reports should include a list of missing, changed and identical files (see Fig. 1). The result file should be stored as a reference file. The verification program will identify whether any file has incidentally been deleted. Therefore, whenever there is a problem with the software, the actual installation should first be checked against this reference file.

A new reference file should be generated whenever the user adds software that has been authorized. An example would be Macros to customize the system. In this way, it can always be checked whether Macros have been changed or added that are not authorized.

The generation and signing off of the installation report should mark the completion of the installation. In the pharmaceutical industry this is referred to as the installation qualification (IQ) document. The document should be signed by the user's representative if the IQ was done by the user, and by the vendor's and the user's representative if the IQ was done by the vendor.

File name	File Description
Missing files	
l\instrmnt.ini	Initialization
repstyle\library.mac	Macro
l\verify\default.val\integ.reg	Register
helpenu\hpsc6a00.hlp	Help
Changed files	
core\800\eevempt.ini	Initialization
core\800\eevtool.ini	Initialization
Identical files	
apg_top.exe	HP APG DataComn
apgdde.dll	HELP

Fig. 1 Installation verification report

Operational qualification and acceptance testing

After the installation of hardware and software, an operational test should follow, a process which is referred to as operational qualification (OQ). For the qualification of computer systems, the term "acceptance" is also used, which can be identical to or be a reduced set of the OQ. The purpose of OQ is to demonstrate that the system's hardware and software operate "as intended" in the user's environment.

Correct functioning of software loaded on a computer system should be checked in the user's laboratory under typical operating conditions within its intended ranges. Tests should be developed that execute the most important software functions. The type and number of tests depend on the complexity of software and on the tests that have already been done during system development. For example, many more tests should be done if there is no evidence of extensive testing done during development. For details see [1].

In an analytical laboratory we can typically find computer systems as shown in Fig. 2.

1. Integrated computerized analytical systems, for example chromatographic systems with computer software for instrument control, data acquisition and data evaluation. Data are printed and electronically stored. Sometimes these computer systems employ spreadsheet programs or user-contributed Macros for customized data evaluation.

2. Networked computer systems with multiple computers and peripherals. Examples are servers for common printing and data storage or client/server networks where the operating system and application software are loaded on the server and can be executed on the client computer. Another example is a laboratory information management system (LIMS) for collection and management of data from multiple computers.

All computers may also employ office software, such as spreadsheets and word processing programs. These programs can operate with manual or automated data entry with on-line connection to the analytical data acquisition and evaluation system.

Testing procedures for categories one and two are very different. In both cases test cases should be developed and acceptance criteria should be defined. The

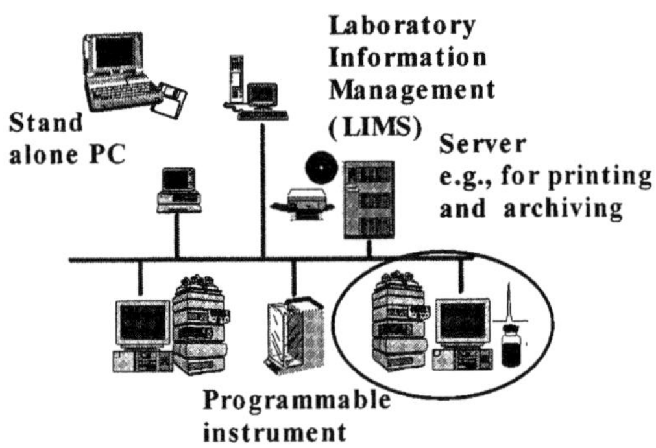

Fig. 2 Examples of computer systems in analytical laboratories

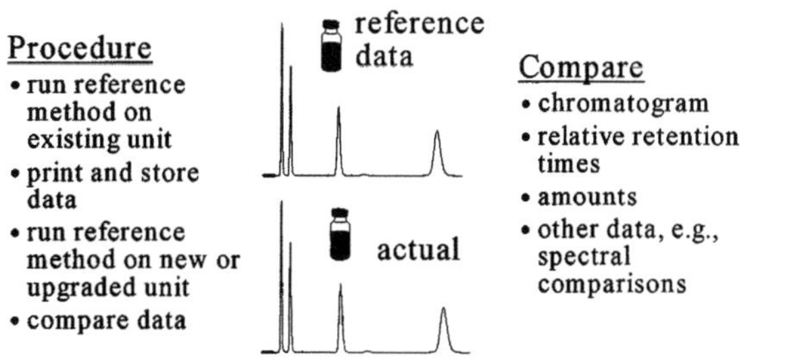

Fig. 3 Testing of computer hardware and software as part of an integrated chromatographic system

most efficient test for integrated analytical system is to analyze a well-characterized chemical standard or sample (see Fig. 3). During this test, key functions of the software are executed, for example, in the case of chromatography

- instrument control
- data acquisition
- peak integration
- quantitation
- file storage
- file retrieval
- printing

If the system has a spectral detector, spectral evaluation such as peak purity and compound identity should be part of the test. Additional tests should be developed if there are other software functions used in routine analysis, but they are not part of the sample or standard analysis test. If the system is used over a wide concentration range with multiple calibration points, the tests should be run over many concentrations to verify correct function of the calibration curve.

Most software functions of a computerized chromatographic system can also be tested by using well-characterized data files without injecting a test sample. The advantage is that less time is required for the test; the results concept has been described in great detail in [3] and is summarized in Table 4. The procedure is very useful after updating the computer system, for example after adding more internal memory or when changing to a new operating system. The test procedure is very generic and can also be used to test and verify the correct functions of other software packages.

Well-characterized test chromatograms and spectra derived from standards or real samples are stored on disk as a master file. Chromatograms and spectra may be supplied by the vendor as part of the software package. These vendor-supplied chromatograms and spectra are only useful if they reflect the user's way of working, otherwise test chromatograms and spectra should be recorded by the user. This *master data file* passes through normal data evaluation, from spectral evaluation and integration to report generation. Results are stored on the hard disk. Exactly the same results should always be obtained when using the same data file and method for testing purposes. If the chromatographic software is used for different methods, the test should be for different methods. For example one test can be set up for assay and an other for impurity tests.

Table 4 Qualification process of chromatographic software

Generation of master data

1. Generate one or more master chromatograms (the chromatograms should reflect typical samples)
2. In case the method uses spectral data, generate master spectra with spectral libraries.
3. Develop integration method, calibration method and procedure for spectral evaluation, for example, peak purity or/and identity checks.
4. Generate and print out master result(s).
5. Save results on paper and electronically.

Verification

1. Select master chromatogram (with spectra).
2. Select master method for data processing (integration, quantitation, spectral evaluation, etc.).
3. Run test manually or automatically. Automation is preferred because it is faster and has less chances for errors.
4. Compare test results with master data. Again, this can be done automatically if such software is built into the system.
5. Print and archive results.

Preferably, tests and documentation of results should be done automatically, always using the same set of test files. In this way users are encouraged to perform the tests more frequently, and user-specific errors are eliminated. In some cases, vendors provide test files and automated test routines for verification of a computer system's performance in the user's laboratory. Needless to say, the correct functioning of this software should also be verified. This can easily be done by changing the method or data file and rerunning the test. The report should indicate an error. If such automated verification software is not available, the execution of the tests and verification of actual results with prerecorded results can be done manually.

Successful execution of such procedure ensures that

- the actual version of the application software works correctly for the tested functions
- executed program and data files are loaded correctly on the hard disk
- the actual computer hardware is compatible with the software

the actual version of the operating system and user interface software is compatible with the application software.

For a networked computer system, OQ can mean, for example, verifying correct communication between the computers and the peripherals. Data sets should be developed and input at one part of the network. The output at some other part should be compared with the

input. For example, if a server is used to secure and archive data from a chromatographic data station, results should be printed on
1. the chromatographic data system
2. the server after storage and retrieval of the files.

The results should be compared, either manually or automatically. If the network links to other in-house systems, correct function of the linkage should be verified using well-characterized data sets.

Some important points should be considered for the OQ of software and computer systems:

1. A validation team should be formed consisting of analytical experts from the laboratories affected, computer experts from IT departments and validation experts.

2. A validation plan should be developed that describes the purpose of the system including subsystems, responsible persons, test philosophy and a schedule for testing.

3. The intended use and functions of the network and all subsystems should be defined. For subsystems and some core functions of the network the vendor should provide a list with detailed functional specifications. From these specifications, the user can derive the functions the systems will use in the user's laboratory.

4. Tests should be developed for each subsystem, and each subsystem should be validated. Again, the vendor should provide validated software to automatically execute these tests. An example was shown above for chromatographic data systems.

5. Acceptance criteria should be specified before the tests start.

6. For multi-user system, some tests should be done while the maximum number of users are using the system.

7. All or at least some tests should be done under conditions of maximum data flow.

8. When there is a change to the system, the validation team should evaluate the possible impact of the change to other parts of the system. Based on this evaluation, a test plan should be developed that executes either all or part of the tests as specified in 3 above.

At the end of the OQ, documentation should be available or developed that should includes a validation protocol with
- the description, intended use and unique identification of equipment
- functional specifications
- test protocols with test items, acceptance criteria, actual test results, date and time when tests have been performed, and people who performed the tests with names and signatures
- summary of results and a statement on the validation status

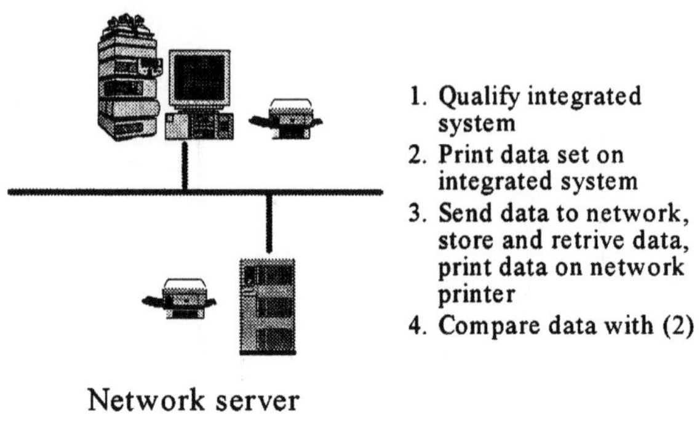

Fig. 4 Qualification of a network server for data storage and printing

Conclusion

Installation and operational qualification (acceptance testing) should follow standardized procedures and the results should be documented. Installation qualification consists of checking the instrument and documentation for completeness. For complex software, the completeness and integrity of the installed program and data files should be checked.

During operational qualification, the software and computer system is verified against the functional specifications. For integrated analytical computerized systems the tests mainly consist of running well-characterized test samples, and the communication to the equipment hardware should be checked as well. For testing the computer system only, laboratory-specific data sets and evaluation procedures can be used.

For networked systems, each individual subsystem should be validated. After successful validation of the subsystems, network functions should be validated. The tests should be defined by a validation team that consists of expert members from various departments.

References

1. Huber L, Wiederoder H (1997) Accred Qual Assur 2:360–366
2. Huber L (1998) Accred Qual Assur 3:140–144
3. Huber L (1995) Validation of computerized analytical systems. Interpharm, Buffalo Grove, Ill. ISBN 0-935184-75-9, Hewlett-Packard Part number: 5959–3879

Accred Qual Assur (1998) 3:317–321

L. Huber

Qualification and validation of software and computer systems in laboratories

Part 4. Evaluation and validation of existing systems

Received: 30 April 1998
Accepted: 2 June 1998

L. Huber (✉)
Hewlett-Packard Waldbronn,
P.O. Box 1280,
D-76337 Waldbronn, Germany
Fax: +49-7802-981948
e-mail: ludwig_huber@hp.com
http://www.t-online.de/home/HUBERL/validat.htm

Abstract Existing software and computer systems in laboratories require retrospective evaluation and validation if their initial validation was not formally documented. The key steps in this process are similar to those for the validation of new software and systems: user requirements and system specification, formal qualification, and procedures to ensure ongoing performance during routine operation. The main difference is that frequently qualification of an existing system is based primarily on reliable operation and proof of performance in the past rather than on qualification during development and installation.

Key words Validation · Qualification · Computers · Analytical laboratories · Existing systems

Introduction

Software and computer systems used in analytical laboratories that must comply with good laboratory practice (GLP) and good manufacturing practice (GMP) regulations require formal validation to confirm that they meet the criteria for intended use. Both new and existing systems, regardless of age, must be validated for their suitability for indented use. Whereas the first three articles in this series [1–3] focused on validation during development, vendor qualification, and installation and operational qualification of new systems, this article describes the procedures for systems that have been purchased, installed, and used prior to formal validation and that must be newly documented.

The Organization for Economic Cooperation and Development (OECD) GLP consensus paper No. 10 [4] contains the following paragraph on retrospective evaluation: "There will be systems where the need for GLP compliance was not foreseen or not specified. Where this occurs there should be documented justification for use of the systems; this should involve a retrospective evaluation to assess suitability".

Recommendations for the retrospective evaluation of computer systems, notably in pharmaceutical manufacturing, have been discussed in the literature. For example, Agalloco [5] has proposed guidelines for the validation of existing computer systems and Hambloch [6] has suggested a strategy for the retrospective evaluation of computer systems in compliance with GLP and GMP. Representatives from the United States Food and Drug Administration as well as from the pharmaceutical industry gave their view on validation of existing legacy systems in a recent publication [7]. However, although this literature provides recommendations on how to retrospectively evaluate computer systems in general, there are no guidelines on the evaluation and validation of existing computerized analytical systems. In this paper, therefore, I attempt to explain how to develop and implement a strategy for the retrospective evaluation and formal validation of existing computerized analytical systems.

The entire process comprises three steps:

1. Identify all software and computerized systems in an analytical laboratory and assess their need for validation. Create a priority list and time schedule for implementation. The priority list can be based on a risk/exposure analysis.
2. Develop a standard operating procedure (SOP) and templates for implementation.

3. Develop a validation plan and evaluate individual systems according to the priority list from step 1 and the SOP from step 2.

Step 1 can be accomplished by defining the environment in which the system is used, by determining the type of data generated by the system, and by estimating the risk of generating inaccurate data or of loosing data. For example, all systems used in a GLP or GMP environment to generate critical data should be validated. Such systems can include computers controlling analytical systems, such as gas chromatographs, spectrophotometers, mass spectrometers but also spreadsheets, Macroprograms and servers for data collection, data archiving and print-out.

Step 2 is more difficult and can be quite complex. This paper therefore focuses on this second step in that it recommends a procedure for the retrospective evaluation. The specific limitations and advantages of older systems compared with new systems are briefly discussed.

Specific characteristics of older systems

Although most programmers and developers of computer systems test products prior to release, almost none of the older systems found in laboratories have been validated and documented according to the newest regulations. For example, the software may not have been developed in accordance with documented quality systems or with the most recent product life cycle guidelines. Testing may have never been formalized, acceptance criteria may not have been specified, the complete test documentation may not have been archived or the required signatures for authorization may not have been collected because at the time there was no real need to do so. As a result, vendors of software and computerized systems often cannot provide retrospective evidence that the software was properly verified and qualified during development. Even if initial tests have been performed and documented at the user's site, almost all computerized systems undergo changes during their lifetime that often are not recorded. Based on these observations, it is easy to conclude that older systems never can attain validation status.

Fortunately, however, existing computerized systems have an advantage over new systems: experience gained over time. Most computer systems in analytical laboratories have been tested in one way another, for example, the vendors functional and performance specifications may have been verified when the system was installed, or the complete system may have been tested for suitability on a day-by-day basis. In the validation process, the evaluator can use this experience and test results to review and assess the quality of analytical results from computer systems. Such a review and its thorough documentation may serve as sufficient evidence that the system has performed and continues to perform according to its initially specified intended use. Because not all the steps required for official validation may be executed and because the final system status can be determined based on a historical review, the process is known as retrospective evaluation.

Procedure for retrospective evaluation

The exact course of a retrospective evaluation depends on the complexity of the system and its use. Therefore, the effort involved will vary from system to system. However, in principle, each evaluation should follow the same procedure. A generic SOP and validation plan should be developed to document the individual steps, which are listed in Table 1. For smaller projects, some steps may be excluded, but the reason for their exclusion should be documented in the protocol for each evaluation.

The first step in the evaluation process is to define and document the current system use and user requirement specifications. If the system will be changed in the foreseeable future, any resulting changes in the intended use of the system should be described as well. The definition should include a list of system functions, operational parameters and performance limits. For chromatography software required functions may include instrument control, data acquisition, peak integration through quantitation, file storage and retrieval and print-out of methods and data. If the system also includes spectrophotometric detectors, the functions for spectral evaluation should be specified as well. Table 2 lists items that should be included in the system documentation.

After specifications have been set, general information on the system should be gathered. The main sources of such information are system users within the organization. Information can also be exchanged among users from different laboratories. Key issues in this phase are the extent of software use and the number of problems reported. This information is not only useful to test past and current performance but can also give some hints on potential problems if the intended use may change in the future.

The next step is to collect information and documentation on the history of the specific system under evaluation. This information should include not only the type and frequency of initial and ongoing testing but also the dates and installation procedure of both software and hardware updates to the system.

Finally, the documentation should be evaluated in relation to the user requirement specifications, functional specifications and performance limits previously defined. There should be a direct link between the tests

Table 1 Recommended steps for retrospective evaluation and validation (from Ref. [8])

1. Describe and define the system.

(a) Summarize intended use, for example, automated qualitative and quantitative analysis of drug impurities.
(b) Specify user requirements for the system.
(c) Define specifications and operating limits of hardware modules, e.g. baseline noise of an high-performance liquid chromatography (HPLC) detector or flow rate precision
(d) List system functions and operating parameters and limits as used for current applications and for intended future use, for example: simultaneous control and data acquisition from two HPLC systems, peak integration, qantitation using an external standard method, peak purity checks using a diode-array detector or statistical evaluation of peak retention times and areas.
(e) List equipment hardware (e.g. spectrophotometer):
- asset number
- merchandising number or name
- manufacturer's name, address and phone number
- hardware serial number and firmware revision number
- date received in the laboratory, date put into operation
- location
- dates of any changes, for example software updates

(f) List computer and peripherals:
- manufacturer's name
- model and serial number
- internal memory
- graphics adapter and monitor
- hard disk
- interfaces
- printer

(g) List all software programs:
- operating system with product and version numbers
- standard applications software
- user-developed applications software
- databases, spreadsheet and word processing programs

(h) List accessories (cables, spare parts, etc).
(i) Provide system drawings.
(j) Define network configurations.

2. Gather all available information and documentation on the standard system.

(a) Information on where, how and how often the same system is used within and outside the organization and reports from users on its performance and reliability (should include both positive and negative comments).
(b) Validation certificates from vendors for purchased systems.
(c) Internat reports for systems developed in-house (development process, source code, quality assurance principles, test procedures and results).
(d) The vendor's performance specifications of equipment hardware, e.g. baseline noise of a UV-visible HPLC detector.
(e) The vendor's functional specifications for software, e.g. peak integration, quantification using external and internal standard methods.
(f) Formulae used for calculations, e.g. system linearity.
(g) User manuals and SOPs on operation, testing, calibration and maintenance.
(h) Records on system updates.
(i) Internal and external audit reports.

Table 1 (continued)

3. Collect information on the history of the specific system.

(a) Installation reports.
(b) Reports on initial testing, e.g. tests to verify that the system meets vendor equipment hardware specifications.
(c) System failure and repair reports.
(d) Equipment hardware maintenance logs, e.g. to record the change of an HPLC pump seal.
(e) Equipment hardware updates, for example, when a variable wavelength detector was replaced by a diode-array detector.
(f) Software updates, for example, when a new revision of the operating system was installed.
(g) Hardware calibration records.
(h) Results on performance monitoring (e.g. results of software tests using test data sets and of system suitability tests or quality control charts using well-characterized standards or control samples).
(i) System feedback reports to the programmer or vendor and responses from the programmer or vendor.

4. Qualify the system.

(a) Check whether the documentation listed under items 2 and 3 is complete and up-to-date, for example, does the revision of the existing user manual comply with the currently installed firmware and software revisions?
(b) Evaluate type of tests and test results and compare the results with user requirements, functional and performance specifications as defined in items 1 b–d. For examples, check if the tests cover only normal operation ranges or also upper- and lower-operational limits.
(c) Assess whether the equipment performs as expected and formally assign the system formal validated status.
(d) If there are not enough tests to prove that the system performs as intended, such tests should be developed, executed and the results documented.

5. Update documentation and develop and implement a plan for maintaining system performance.

(a) Update system description, specifications, drawings, appropriate SOPs, and user manual.
(b) Develop a preventive maintenance plan.
(c) Develop a calibration schedule.
(d) Establish procedures and schedule for performance qualification.
(e) Develop error recording, reporting, and a remedial action plan.

and each user requirement, functional and performance specification as listed in the first step (see Table 3). The tests should confirm that the system operates as intended over all specified ranges for each function. Tests should have been performed not only at typical operating ranges but also at the operational limits. For example, if the system runs multiple applications simultaneously, tests should verify system performance under such conditions.

If the tests yield enough documentation to confirm system performance according to specifications over all operational ranges, the system can be formally released as a validated system. A plan should then be developed

Table 2 Documentation of a computer for retrospective evaluation

System I.D.
Application
Location
Installation date
Installed with current configuration since (date)
Computer hardware
Monitor
Printer
Connected equipment hardware
Network
Operating software
Application software
System functions
Major repairs
Comments

Table 3 Test matrix

Function (user requirement)	Evidence of correct performance
Function 1	yes, see protocol AW1
Function 2	yes, see protocol AW2
Function 3	yes, see protocol AX77
Function 4	no, tests must be defined

to ensure that system performance is maintained in the future. Such a plan should include preventative maintenance steps and calibration and test schedules as well as a change control procedure.

If insufficient tests are available to verify proper system performance, a test plan should be developed to execute missing functions and to test functions over missing operational ranges. For example, if a UV-visible diode array is used in an High-performance liquid chromatography (HPLC) system to evaluate peak purity or to confirm peak identity, a test plan should include tests for these parameters if the original tests did not cover these functions. In such situations, serious consideration should be given to updating the system software. It often costs less to upgrade or replace the software than to validate a new system. In some cases, however, software updates can be rather expensive because the hardware typically must be updated or replaced completely as well.

Problems also arise if the new tests detect an error in the software. For example, an error in the algorithm for data evaluation will yield incorrect final data. Therefore, the SOP should also contain steps to be taken if the previously generated data is incorrect, such as notification procedures.

After evaluation and validation, the following minimum documentation should be available:

- a standard operating procedure for implementation and validation protocol
- user requirement and functional specifications
- a description of the system hardware and software, including all functions
- historical logs of the hardware with system failure reports, maintenance logs and records, and calibration records
- test data demonstrating that the system performs according to user requirements and functional specifications
- SOPs and schedules for preventive maintenance and ongoing performance testing
- error recording, reporting, and remedial action plan
- a formal statement with appropriate signatures that the system is formally validated
- up-to-date user information documentation such as operating and reference manuals
- validation report

Conclusions

All software and computer systems used in accredited and GLP- and GMP-regulated analytical laboratories for the generation and analysis of critical data should be retrospectively evaluated and formally validated. As a first step, the user should list all systems in the laboratory, assess the need for validation for each system and develop an implementation plan. For validation of individual systems, the validation plan should describe user requirements and specific validation steps. It should also list the person or persons responsible for each step as well as the criteria that must be met to achieve validated status.

Retrospective evaluation begins with the collection of system information such as test results, logs of failures and repairs, and records of system changes. The user must them determine whether enough tests have been performed to verify proper system performance as described in the user requirements document. If there is not enough evidence that the system performs as intended, additional tests should be developed and executed. If the system passes all the requirements, it should be treated as a new system.

References

1. Huber L (1997) Accred Qual Assur 2:360–366
2. Huber L (1998) Accred Qual Assur 3:2–5
3. Huber L (1998) Accred Qual Assur 3:140–144
4. OECD (1995) Application of the principles of GLP to computerized systems. (Series on principles of good laboratory practice and compliance monitoring, No. 10, GLP, consensus document environment monograph No. 116) Paris
5. Agalloco J (1987) Farm Tech , 38–40
6. Hambloch H (1994) Existing computer systems: a practical approach to retrospective evaluation. In: Computer validation practices. Interpharm, Buffalo Grove, Ill., USA, p 93-112
7. The Gold Sheet, Quality Control Reports, vol 30, No. 7, July 1996. Chevy Chase, Md., USA, pp 3, 9, 10
8. Huber L (1995) Validation of computerized analytical systems. Interpharm, Buffalo Grove, Ill., USA

Accred Qual Assur (1997) 2:140–145

Reports and notes on experiences with quality assurance, validation and accreditation

S. Seno
S. Ohtake
H. Kohno

Analytical validation in practice at a quality control laboratory in the Japanese pharmaceutical industry

Received: 4 October 1996
Accepted: 5 December 1996

S. Seno (✉) · S. Ohtake · H. Kohno
Quality Assurance Department, Sandoz Pharmaceuticals Ltd., No. 1 Takeno, Kawagoe City, Saitama, Japan

Abstract Analytical validation is required as the basis for any evaluation activities during manufacturing process validation, cleaning validation and validation of the testing method itself in the pharmaceutical industry according to good manufacturing practice (GMP) rules and guidelines. Validation of analytical methods and procedures in a quality control (QC) laboratory is implemented mainly at the time of transfer or introduction of the methods developed by the analytical development laboratory within group companies or elsewhere. However, it is sometimes necessary to develop a new or improved method of analysis for the QC laboratory's own use. In the first part of this report, a general description of analytical validation of the high performance liquid chromatography (HPLC) method including preparation of documents is presented based on the experience in our QC laboratory. A typical example of method validation of robotic analysis system is then cited. Finally the merits and demerits of these analytical validations for QC laboratories are summarized. The authors emphasize the importance of analytical validation and the responsibility of QC laboratory management for the effective design and implementation of validation activities.

Key words Analytical method and procedure · Validation · Quality control laboratory · Documentation · Robotic analysis

Introduction

In the area of pharmaceutical industries in most countries, validation of manufacturing processes and supporting functions is the fundamental issue for the quality assurance of products. Analytical methods and procedures are associated with most evaluation activities in this field. Analytical validation is an essential prerequisite for any evaluation work during process validation, cleaning validation and testing method validation itself according to GMP regulations, rules and guidelines [1–3].

In most quality control (QC) laboratories in this field, there are a number of types of analytical instruments and equipment in practical use. These must be suitable for the intended use in analytical work. Installation qualification (IQ) and operational qualification (OQ) are required before practical use. Calibration of the instruments according to the preset schedule and preparation of the standard operating procedures (SOPs) must be implemented and records must be kept.

Validation of analytical methods and procedures is one of the important duties of development laboratories in the R and D division of major pharmaceutical companies. Validation of the testing method in a QC laboratory is implemented mainly at the time of transfer or introduction of the method from the development laboratory of the group companies or from outside. However, it often becomes necessary to develop a new or improved method of analysis for the QC laboratories' own use.

Analytical validation also necessitates documentation prepared and approved for record keeping. It takes resources and time to collect data and elaborate validation parameters to establish the method and to compile documentation. It is the responsibility of the QC laboratory management to design and implement these validation activities efficiently before proceeding with routine analysis.

Definitions

The analytical method or procedure is a detailed description of the steps necessary to perform each analytical test. This may include (but is not limited to) preparation of the sample, reference standard and reagent, use of apparatus, generation of the calibration curve, use of the formulae for calculation, etc.

Validation of analytical procedure is the process of confirming that the analytical procedure employed for a test of pharmaceuticals is suitable for its intended use [4].

Validation of an analytical method is the process by which it is established, by laboratory studies, that performance characteristics of the method meet the requirement for the intended analytical applications [5].

The test methods, which are used for assessing compliance of pharmaceutical products with established specifications, must meet proper standards of accuracy and reliability [2].

Analytical method validation

As required by regulatory authorities [1–3, 6] and International Conference on Harmonisation (ICH) guidelines [7], analytical method validation is especially important in establishing the assay methods and procedures of quantitative or semi-quantitative measurement of target substances or compounds. Among the specification items for drug products, assay, content uniformity, and dissolution are typical of those which require almost full analytical validation. The purity test, related substances test, and residual solvent test may be categorized as semi-quantitative assays which require somewhat reduced validation work. For the manufacturing validation, multi-point sample assay must be performed to verify the prescribed homogeneous distribution of an active drug substance and uniformity throughout a batch of the product. For cleaning validation, a semi-quantitative assay of low level active substance and detergent used for cleaning is necessary. Various analytical techniques and methods can be used to meet the requirements. A typical and versatile analytical technique for this purpose is high performance liquid chromatography (HPLC).

Steps in the analytical validation of the HPLC method

Preliminary work

Availability of development manual

Documentation of policies, fundamental principles, and procedures for development are prerequisites for the implementation of validation activities and should be available.

Selection of the analytical technique and instrument

The most important factor for selection of the analytical technique and instrument is the objective of the analysis. The method must meet the requirements of guidelines issued by the regulatory authorities where the method is intended for compliance purposes.

Features of the sample and availability of reference materials are also key factors in determining an analytical procedure and detection method, which then relate to various analytical validation characteristics and parameters.

Cost effectiveness is a pressing factor even in QC laboratories, while at the sametime quality standards must be kept as high as necessary. Time and manpower-saving methods must be developed and applied in routine operations in the QC laboratory of the pharmaceutical company. There are many other limiting factors for establishing the procedure. The HPLC method is nowadays the most widely used versatile method in the pharmaceutical area.

IQ and OQ of the HPLC instrument

Elements of IQ

IQ is the verification based on installation documents that indicate that the instrument meets the requirements of the system and that all components of the system have been delivered, ready for installation in accordance with the standards and specifications. The areas designated for equipment installation must have adequate space and appropriate utilities. A check list or a table format is often prepared to facilitate IQ for approval and for record keeping.

Elements of OQ

OQ is the verification that the installed system operates as prescribed for the intended purpose in accordance with the manufacturer's specification as well as the user company's standard. OQ includes calibration and verification of the unit components of the system under operating conditions. SOPs must be developed at this stage.

The UV detector, fluorescence detector, electrolytic detector, refractive index detector, and other detection systems must be checked for repeatability of response signal, sensitivity, and stability.

The pump must give reproducibility of flow rate and stability at maximum and minimum flow rates.

The auto-sampler must give reproducibility of injection volume and stability of controlled temperature.

The oven must have accuracy and stability of controlled temperature.

The integrator/data processor/PC or micro-computer must give reliability in terms of computer validation and standard output versus input.

Performance qualification (PQ) of an HPLC system

Analytical validation documentation

Validation documentation, consisting of a protocol, test data, and reports, must be reviewed and approved by responsible personnel. Examples of the contents in a protocol and a report are shown below:

1. Protocol
   Method name: Validation number: Date:
   Objective:
   Summary of method:
   Specific requirements:
   Sample and analyte:
   Reference standard:
   Characteristics to be validated (see Table 1):
   Detailed description of method/procedure:
   Data sheet/format for each validation characteristic:
   Revalidation and reason:
   Prepared by: Date:
   Reviewed by: Date:
   Approved by: Date:
2. Report
   Title/method name: Validation number: Date:
   Protocol:
   Data sheets and raw data including chromatograms:
   Results versus acceptance criteria:
   Evaluation and comments:
   Conclusion:
   Revalidation schedule
   Preprepared by: Date:
   Reviewed by: Date:
   Approved by: Date:

System suitability test (SST)

The SST is a single overall test of system function, and is used to verify that the selectivity and repeatability of a chromatographic system is adequate for the analysis required. The SST must be performed before the analysis. The method/procedure should describe the frequency of the SST.

Table 1 Characteristics to be validated in the HPLC method (example of format)

Characteristic	Required?	Acceptance criteria	Remarks/ Other criteria
Accuracy/trueness	Y, N	Recovery 98–102% (individual) with 80, 100, 120% spiked sample	
Precision			
Repeatability	Y, N	RSD <2%	
Intermediate precision	Y, N	RSD <2%	
Specificity/selectivity	Y, N	No interference	
Detection limit	Y, N	S/N >2 or 3	
Quantitation limit	Y, N	S/N >10	
Linearity	Y, N	Correlation coefficient r>0.999	
Range	Y, N	80–120% or QL–120% or 200%	
Stability of sample solution	Y, N	>24 h or >12 h	

The terms are as defined in [4] and [5].

1. Selectivity: system selectivity chromatogram must be attached.
   Retention time (t)
   Resolution (Rs)
   Selectivity (α)
   Number of theoretieal plates (N)
   Symmetry factor (S) or Tailing factor (T)
2. Repeatability of peak area or peak height
   In assay: using 100% reference solution with at least three injections, for validation purpose, not less than six injections, RSD <1%
   In purity test: using 100% reference solution with at least three injections, RSD <1%
   using 1% reference solution with at least three injections, RSD <2%.

Typical example of analytical validation in the QC laboratory: method validation of a robotic analysis system for ketotifen 1 mg capsules and tizanidine hydrochloride 1 mg tablets

A Zymate II Pye system for automated analysis by the Zymark Corporation (USA) with on-line HPLC has been set up and used for content uniformity tests and assays, including process validation, for a variety of drug products in our QC laboratory since January 1992. The robotic system automatically performs the following operations: preparation of sample and reference standard solutions, injection into HPLC, HPLC analyses, and generation of reports.

The robotic operation procedures were designed to simulate the manual ones (registered test methods) and were slightly modified to fit the robotic system. HPLC parameters and conditions are the same as those used in the routine manual methods.

Configuration of Zymate II Robotic System

Flow diagram of basic robotic assay procedure

Major validation elements

Master laboratory station (MLS) for pipetting and dilution with 3 (A, B, C) 10 ml syringes

Accuracy
Repeatability of delivered volume

Disposable 5 ml pipette tips for pipetting 1 ml, 2 ml and 4 ml of solvents

Accuracy
Repeatability of delivered volume

Filter cartridges

Recovery (adsorption loss)

Content uniformity test of ketotifen 1 mg capsules and tizanidine hydrochloride 1 mg tablets

Comparison of manual analysis and robotic analysis with the same procedure.
Comparison with registered method.

Validation results

Master laboratory station (MLS)

Accuracy and repeatability of delivering 2 ml, 5 ml and 10 ml of three different solvents (100% water, 50% water/methanol, 100% methanol) were checked by weighing the delivered solvents. Accuracy was within the acceptance limit of 98.0% of the theoretical values and the RSD was less than 2% for all 2 ml, 5 ml, 10 ml deliveries.

Disposable 5 ml pipette tips

Deliveries of preset volume of 1 ml, 2 ml, and 4 ml of the same solvents used for MLS were checked by weighing. For water and 50% water/methanol, three levels of delivery were within acceptance limits and showed no difference from manual pipetting. However, 50% water/methanol and 100% methanol gave smaller volumes, and out of limit for 1 ml and 2 ml deliveries.

Filter cartridges

There was no detectable loss of ketotifene and tizanidine hydrochloride in filter cartridges.

Content uniformity test

Results are shown in Table 2.

There was no significant difference in the results obtained by manual operation, robotic operation and the registered method. All the data were within the acceptance limits.

The validation of the robotic method was successfully completed by finalizing the validation report and approving the documentation by the management of the Quality Assurance Department.

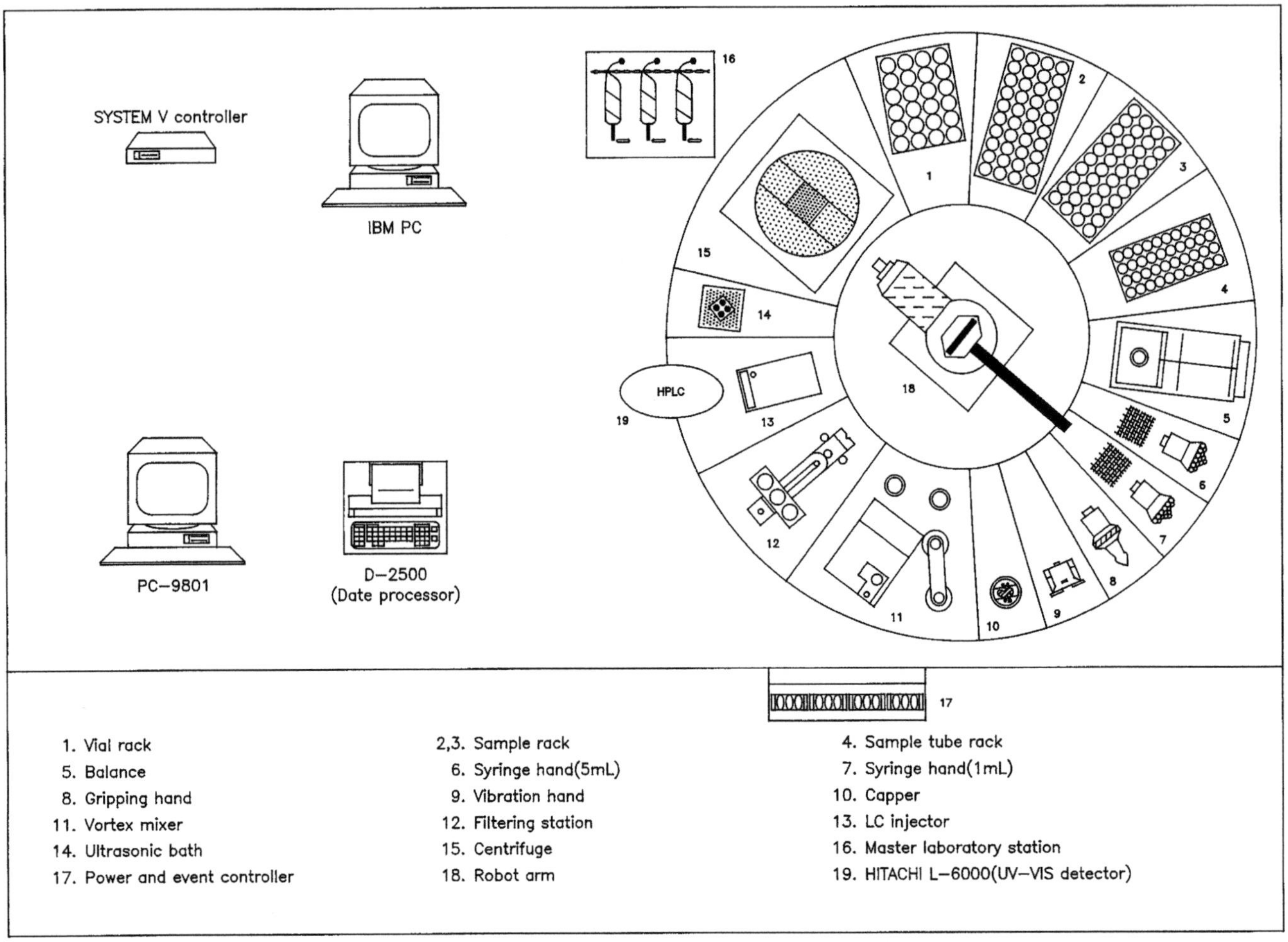

Fig. 1 Configration of Zymate® II Robotic System

Fig. 2 Flow diagram of basic robotic procedure for content uniformity assay.

Weighing (Mettler balance AE200)
↓
Addition of designated volume of solvent (MLS)
↓
Extraction of active substance (Vortex mixer)
↓
Centrifugation (Centrifuge)
↓
Filtration (Filter cartridge)
↓
Dilution (MLS and syringe hand with 1 ml, 5 ml pipette tips)
↓
Injection into HPLC
↓
HPLC analysis (Hitachi L-6000 with UV-VIS detector)
↓
Data processing (Hitachi L-2500 integrator)
↓
Report generation (NEC PC9801 DX)

Merits and demerits associated with analytical validation in the QC laboratory

Merits

1. Reliability of analytical results is assured through the validation.
2. The clear objective and performance capability of the method can be confirmed by all the staff who apply the method.
3. Awareness of the importance of predetermined protocols for the validation work and motivation for the improvement can be enhanced.
4. Analytical validation gives a good opportunity for training the QC staff.

Demerits

1. Costs for validation are continually increasing.
2. Designing and scheduling of validation need experienced personnel.

Table 2 Comparison of the manual, robotic, and registered methods for content uniformity (*Manual,* Unit operations in the robotic method were performed manually; *Registered,* Manual assay method using HPLC cited in the registration document)

Product	Lot No.	Percentage of nominal content found by various methods (RSD)		
		Manual	Robotic	Registered
Ketotifen 1-mg capsules	66562	101.0 (0.9)	101.6 (0.4)	101.1 (1.1)
	66564	101.8 (2.4)	102.3 (0.3)	100.2 (2.4)
	66569	100.4 (1.5)	101.2 (0.6)	101.0 (1.5)
Tizanidine hydrochloride 1-mg tablets	66563	99.5 (1.0)	99.3 (0.4)	99.5 (0.3)
	66565	99.5 (1.1)	99.3 (0.6)	98.3 (0.5)
	66570	99.7 (1.5)	99.6 (1.6)	99.5 (0.9)

Conclusion

Reliability of analytical results generated in a QC laboratory depends on the analytical method and procedure used. Unless the method and procedure are validated, the numerical results remain as an estimate and cannot be used for rigorous compliance judgment. The quality of the pharmaceutical product must be assured from the very beginning of the manufacturing activities, including acceptance testing of raw materials, processing operations, in-process controls, and product testing together with established manufacturing processes and validated systems.

Analytical validation is the fundamental element supporting quality assurance activities in pharmaceutical industries. It is the responsibility of QC laboratory management to design and implement these validation activities efficiently and effectively.

References

1. Regulations for Manufacturing Control and Quality Control of Drugs (GMP of Japan), amended in January 1994, Ministry of Health and Welfare, Japan
2. CFR Title 21 Part 211-Current Good Manufacturing Practice for Finished Pharmaceuticals, revised in January 1995, Food and Drug Administration, Department of Health and Human Services, USA
3. Guide to Good Manufacturing Practice for Medicinal Products, The GMP Directives 91/356/EEC and 91/341/EEC (1991), European Commission
4. The Japanese Pharmacopeia, Thirteenth Edition (JP 13) (1995) Ministry of Health and Welfare, Japan
5. The United States Pharmacopeia, Twenty-third Revision (USP 23) (1995) United States Pharmacopeial Convention, Rockville, MD, USA
6. Guidelines for Establishing the Specifications and Analytical Procedures of New Drugs, Notification No. 586 (1994) by the director of the Pharmaceuticals and Cosmetic Division, MHW, Japan
7. ICH Harmonised Tripartite Guideline "Text on Validation of Analytical Procedures", October 1994, and Draft Consensus Guideline "Extension of the ICH Text on Validation of Analytical Procedures", November 1995, by the "ICH Steering Committee, International Conference on Harmonisation of Technical Requirements for Registration of Pharmaceuticals for Human Use

Accred Qual Assur (1998) 3:155–160

Reports and notes on experiences with quality assurance, validation and accreditation

Ricardo J. N. Bettencourt da Silva
M. Filomena G. F. C. Camões
João Seabra e Barros

Validation of the uncertainty evaluation for the determination of metals in solid samples by atomic spectrometry

Received: 3 November 1997
Accepted: 2 January 1998

Presented at: 2nd EURACHEM Workshop on Measurement Uncertainty in Chemical Analysis, Berlin, 29–30 September 1997

R. J. N. Bettencourt da Silva (✉)
M. F. G. F. C. Camões
CECUL, Faculdade de Ciências da Universidade de Lisboa, P-1700 Lisbon, Portugal

J. Seabra e Barros
Instituto Nacional de Engenharia e Tecnologia Industrial, Estrada do Paço do Lumiar, P-1699 Lisbon Codex, Portugal

Abstract Every analytical result should be expressed with some indication of its quality. The uncertainty as defined by Eurachem ("parameter associated with the result of a measurement that characterises the dispersion of the values that could reasonably be attributed to the, ..., quantity subjected to measurement") is a good tool to accomplish this goal in quantitative analysis. Eurachem has produced a guide to the estimation of the uncertainty attached to an analytical result. Indeed, the estimation of the total uncertainty by using uncertainty propagation laws is components-dependent. The estimation of some of those components is based on subjective criteria. The identification of the uncertainty sources and of their importance, for the same method, can vary from analyst to analyst. It is important to develop tools which will support each choice and approximation. In this work, the comparison of an estimated uncertainty with an experimentally assessed one, through a variance test, is performed. This approach is applied to the determination by atomic absorption of manganese in digested samples of lettuce leaves. The total uncertainty estimation is calculated assuming 100% digestion efficiency with negligible uncertainty. This assumption was tested.

Key words Uncertainty · Validation · Quality control · Solid samples · Atomic spectrometry

Introduction

The presentation of an analytical result must be accompanied by some indication of the data quality. This information is essential for the interpretation of the analytical result. The comparison of two results cannot be performed without knowledge of their quality. Eurachem [1] defined uncertainty as the "parameter associated with the result of a measurement that characterises the dispersion of the values that could reasonably be attributed to the, ..., quantity subjected to measurement" and presented it as a tool to describe that quality.

The Eurachem guide for "Quantifying uncertainty in analytical measurement", which is based on the application of the ISO guide [2] to the chemical problem, was observed. ISO aims at the estimation of uncertainty in the most exact possible manner, in order to avoid excess of confidence in overestimated results. The application of these guides turns out to be a powerful tool. The exact estimation of uncertainties is important for the detection of small trends in analytical data. The time and effort used in such estimations can avoid many further doubts concerning observation of legal limits and protects the user of the analytical data from financial losses. The use of uncertainty instead of less informative percentage criteria brings considerable benefits to the daily quality control.

Despite the analyst's experience, some analytical steps like sampling and recovery are of particularly difficult estimation. Mechanisms should be developed to support certain choices or approximations. The comparison of an estimated uncertainty with the experimentally assessed one can be of help.

In this work the Eurachem guide [1] was used for the estimation of uncertainties involved in the determination by electrothermic atomic absorption spectrometry (EAAS) of manganese in digested lettuce leaves. The total uncertainty estimation was calculated assuming a 100% digestion efficiency with negligible uncertainty. The experimental precision was compared with an estimated one for the purpose of validation of the proposed method of evaluation. After this validation the uncertainty estimation was used in an accuracy test and in routine analysis with the support of a spreadsheet programme.

The uncertainty estimation process

The uncertainty estimation can be divided into four steps [1]: (1) specification, (2) identification of uncertainty sources, (3) quantification of uncertainty components, and (4) total uncertainty estimation.

Specification

A dry-base content determination method is proposed, the sample moisture determination being done in parallel. Figure 1 represents the different steps of the analysis. The analytical procedure was developed for laboratory samples. Sampling uncertainties were not considered.

The dry-base correction factor, $f_{corr.}$, is calculated from the weights of the vial (z), vial plus non-dried sample (x) and vial plus dry sample (y)

$$f_{corr.} = 1 - \frac{x-y}{x-z} \quad (1)$$

The sample metal content, M, is obtained from the interpolated concentration in the calibration curve, C_{inter}, the mass of the diluted digested sample, a, and the dilution factor, $f_{dil.}$, (digested sample volume times dilution ratio).

$$M = \frac{C_{Inter.} \times f_{dil.}}{a} \quad (2)$$

The dry-base content, D, is obtained by application of the correction factor, $f_{corr.}$, to the metal content, M.

$$D = f_{corr.} M \quad (3)$$

Identification of uncertainty sources

The uncertainty associated with the determination of $f_{corr.}$ is estimated from the combination of the three involved weighing steps, Fig. 1a.

The uncertainty associated with the sample metal content is estimated from the weighing, dilution and interpolation sources (Fig. 1b). The model used for the calculation of the contribution from the interpolation source assumes negligible standards preparation uncertainty when compared with the instrumental random oscillation [5, 10].

Quantification of the uncertainty components

The quantification of the uncertainty is divided into equally treated operations:

Gravimetric operations

The weighing operations are present in the dry-base correction factor (three) and in the sample metal content (one). Two contributions for the associated uncertainty, $\sigma_{Weighing}$, were studied:

1. Uncertainty associated with the repeatability of the weighing operations, $\sigma^{Balance}_{Repeat.}$, is obtained directly from the standard deviation of successive weighing operations. The corresponding degrees of freedom are the number of replicates minus 1.

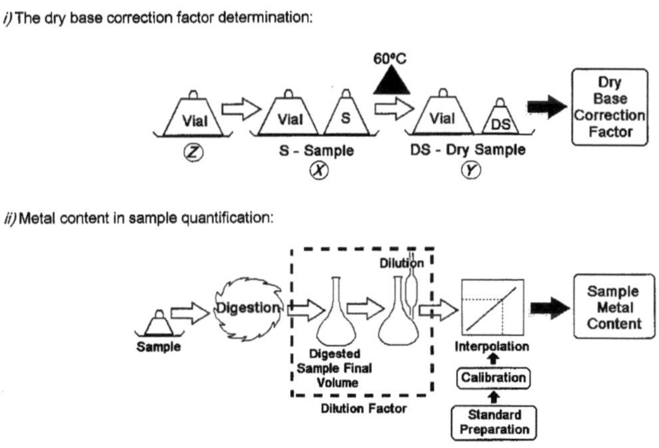

Fig. 1 Proposed method for dry-base metal content determination in lettuce leaves

2. Uncertainty associated with the balance calibration, $\sigma_{\text{Calib.}}^{\text{Balance}}$, defined by

$$\sigma_{\text{Calib.}}^{\text{Balance}} = \frac{2 \times Tolerance}{\sqrt{12}} \quad (4)$$

where the *Tolerance* is obtained from the balance calibration certificate.

The Eurachem guide suggests that when the uncertainty components are described by a confidence interval, $\alpha \pm \beta$, without information on degrees of freedom, the associated uncertainty is $2\beta/\sqrt{12}$, which represents the uncertainty of a 2β amplitude rectangular distribution. These uncertainties are designated type B. The number of degrees of freedom associated with the $\sigma_{\text{Calib.}}^{\text{Balance}}$ type B estimation, $\nu_{\text{Calib.}}^{\text{Balance}}$, is approximately [1] equal to

$$\nu_{\text{Calib.}}^{\text{Balance}} \cong \frac{1}{2}\left[\frac{\sigma_{\text{Calib.}}^{\text{Balance}}}{m_{\text{Calib.}}^{\text{Balance}}}\right]^2 \quad (5)$$

were $m_{\text{Calib.}}^{\text{Balance}}$ is the mass associated with the balance calibration tolerance.

The two uncertainties are then combined

$$\sigma_{\text{Weiging}} = \sqrt{(\sigma_{\text{Calib.}}^{\text{Balance}})^2 + (\sigma_{\text{Repeat.}}^{\text{Balance}})^2} \quad (6)$$

The corresponding degrees of freedom are calculated by the Welch-Satterwaite equation [1–3]. When the pairs (uncertainty, degrees of freedom)

$(\sigma_a, \nu_a); (\sigma_b, \nu_b); (\sigma_c, \nu_c); (\sigma_d, \nu_d); \ldots$

for the quantities $a, b, c, d, \ldots$, in a function $\nabla = f(a, b, c, d, \ldots)$ are taken into account, then the effective number of degrees of freedom associated with ∇, ν_∇, is

$$\nu_\nabla = \frac{\sigma_\nabla^4}{\left(\frac{\partial f}{\partial a}\right)^4 \frac{\sigma_a^4}{\nu_a} + \left(\frac{\partial f}{\partial b}\right)^4 \frac{\sigma_b^4}{\nu_b} + \left(\frac{\partial f}{\partial c}\right)^4 \frac{\sigma_c^4}{\nu_c} + \left(\frac{\partial f}{\partial d}\right)^4 \frac{\sigma_d^4}{\nu_d} + \ldots} \quad (7)$$

The calculation of the uncertainty and of the degrees of freedom associated with the sample weight is by the direct application of Eqs. 4–7. The calculations of the dry-base correction factor are more elaborate.

3. Uncertainty associated with the dry base factor

The dry base correction factor is a function of three weighing operations (Eq. 1). To estimate the uncertainty, $\sigma_{\text{fcorr.}}$, associated with the $f_{\text{corr.}}$, the general equation (Eq. 8) was used [1]

$$\sigma_{f_{\text{corr.}}} = \sqrt{\left(\frac{\partial f_{\text{corr.}}}{\partial x}\right)^2 \sigma_x^2 + \left(\frac{\partial f_{\text{corr.}}}{\partial y}\right)^2 \sigma_y^2 + \left(\frac{\partial f_{\text{corr.}}}{\partial z}\right)^2 \sigma_z^2} \quad (8)$$

It is therefore

$$\sigma_{f_{\text{corr.}}} = \sqrt{\left(\frac{y-z}{(x-z)^2}\right)^2 \sigma_x^2 + \left(-\frac{1}{(x-z)}\right)^2 \sigma_y^2 + \left(\frac{x-y}{(x-z)^2}\right)^2 \sigma_z^2} \quad (9)$$

The values of σ_x, σ_y and σ_z are then calculated as described in the section "Gravimetric operations" above. The number of degrees of freedom is calculated by the Welch-Satterwaite equation (Eq. 7). The application of a spreadsheet program available in the literature simplifies this task [3]. However, the classical approach is more flexible for different experimental configurations or for one or more dilution steps, and is also easily automated.

Volumetric operations

The uncertainties associated with to the volumetric operations were calculated from the combination of two [(1) and (2) below] or three [(1), (2) and (3) below] components:

1. Uncertainty associated with volume calibrations, $\sigma_{\text{Calib.}}^{\text{Vol.}}$

$$\sigma_{\text{Calib.}}^{\text{Vol.}} = \frac{2 \times Tolerance}{\sqrt{12}} \quad (10)$$

where the information on this tolerance is normally available with the instrument in the form: volumetric instrument volume ± tolerance. This type B uncertainty estimation has the same treatment as the one reported in Eq. 5 for the degrees of freedom

$$\nu_{\text{Calib.}}^{\text{Vol.}} \cong \frac{1}{2}\left[\frac{\sigma_{\text{Calib.}}^{\text{Vol.}}}{V}\right]^2 \quad (11)$$

where $\nu_{\text{Calib.}}^{\text{Vol.}}$ is the number of degrees of freedom associated with $\sigma_{\text{Calib.}}^{\text{Vol.}}$ for a certain volume V.

2. Uncertainty associated with volume repeatability tests, $\sigma_{\text{Repeat.}}^{\text{Vol.}}$

The $\sigma_{\text{Repeat.}}^{\text{Vol.}}$ and the corresponding degrees of freedom, $\nu_{\text{Repeat.}}^{\text{Vol.}}$, are also extracted directly from the repeatability tests. Such tests consist of successive weighings of water volumes measured by the instrument. The observed standard deviation is a function of the analyst's expertise.

3. Uncertainty associated with the use of volumetric equipment at a temperature different from that of calibration, $\sigma_{\text{Temp.}}^{\text{Vol.}}$

This third component corrects for errors associated with the use of 20 °C calibrated material in 20 ± 3 °C solutions. When two consecutive volumetric operations are performed at the same temperature, as is the case in dilution stages, they become self-corrected for this effect.

The glass instrument expansion coefficient is much smaller than that of the solution. For this reason we have only calculated the latter. For a temperature oscillation of $\Delta T = \pm 3K$ with a 95% significance level and for a volumetric expansion coefficient of pure water of $2.1 \times 10^{-4}\,°C^{-1}$ (our solutions can be treated as pure water because of their low concentrations), the 95% volume confidence interval becomes $V \pm V \times 3 \times 2.1 \times 10^{-4}$. Dividing the expanded uncertainty by the Student t value, $t(\infty, 95\%) = 1.96$, we obtain the temperature effect component uncertainty

$$\sigma^{Vol.}_{Temp.} = \frac{V \times 3 \times 2.1 \times 10^{-4}}{1.96} \quad (12)$$

The number of degrees of freedom due to the temperature effect can also be estimated as for $\nu^{Vol.}_{Calib.}$ (Eq. 11), substituting $\sigma^{Vol.}_{Calib.}$ by $\sigma^{Vol.}_{Temp.}$

These components are then combined to calculate the volume uncertainty, $\sigma_{Vol.}$

$$\sigma_{Vol.} = \sqrt{(\sigma^{Vol.}_{Calib.})^2 + (\sigma^{Vol.}_{Repeat.})^2 + (\sigma^{Vol.}_{Temp.})^2} \quad (13)$$

The number of degrees of freedom associated with $\sigma_{Vol.}$ can also be calculated by the Welch-Satterwaite equation.

4. Uncertainty associated with the dilution factor

Our analytical method has three volumetric steps that can be combined as a dilution factor, $f_{dil.}$, whose uncertainty, $\sigma_{f_{dil.}}$, can easily be estimated by:

$$\frac{\sigma_{f_{dil.}}}{f_{dil.}} = \sqrt{\left(\frac{\sigma^{DSV}_{Vol.}}{V^{DSV}}\right)^2 + \left(\frac{\sigma^{P}_{Vol.}}{V^{P}}\right)^2 + \left(\frac{\sigma^{V}_{Vol.}}{V^{V}}\right)^2} \quad (14)$$

were the DSV, P and V stand respectively for digested solution volume, dilution operation pipette and dilution operation vial; $\sigma_{Vol.}$ and V represent respectively each corresponding volumetric uncertainty and volume. As in the other cases, the degrees of freedom were calculated by the Welch-Satterthwaite equation.

Sample signal interpolation from a calibration curve

The mathematical model used to describe our calibration curve was validated by the Pennincky et al. [4] method. At this stage we proved the good fitting properties of the unweighted linear model to our calibration curve. With this treatment we aimed not only at the accuracy but also at the estimation of more realistic sample signal interpolation uncertainties. These uncertainties were obtained by the application of an ISO international standard [5].

The instrument was calibrated with four standards (0–2–4–6 μg/L for Mn) with three measurement replicates each [4]. Samples and control standard (4 μg/L for Mn) were also measured three times. The control standard was analysed for calibration curve quality control (see "Quality control").

Total uncertainty estimation

The total uncertainty estimation, σ_T, is a function of the dry-base correction factor uncertainty, $\sigma_{f_{corr.}}$, of the uncertainty associated to the analysis sample weighing operation, $\sigma^{Sample}_{Weighing}$, of the dilution factor, $\sigma_{f_{dil.}}$, and of the instrumental calibration interpolated uncertainty, $\sigma_{C_{inter.}}$. These four quantities combine their uncertainties in the equation

$$\frac{\sigma_T}{D} = \sqrt{\left(\frac{\sigma^{Sample}_{Weighing}}{a}\right)^2 + \left(\frac{\sigma_{f_{corr.}}}{f_{corr.}}\right)^2 + \left(\frac{\sigma_{f_{dil.}}}{f_{dil.}}\right)^2 + \left(\frac{\sigma_{C_{Inter.}}}{C_{Inter.}}\right)^2} \quad (15)$$

where D represents the dry-base sample metal content and a has the same meaning as in Eq. 2. The other quantities have already been described.

The expanded uncertainty can then be estimated after the calculation of the effective number of degrees of freedom, df (Eq. 7). Therefore the coverage factor used was the Student t defined for that number and a 95% significance level (t(df, 95%)). The estimated confidence interval is defined by

$$D \pm \sigma_T \cdot t(\mathrm{df}, 95\%) \quad (16)$$

Quality control

Ideally, the readings of the instruments for each sample and for each standard should be random [6–7]. Normally, the instrument software separates the calibration from the sample reading. Although this allows an immediate calculation, it can produce gross errors if the operator does not verify the drift of the instrument response. For this reason, the calibration curves should be tested from time to time by reading a well-known control standard. This standard can also be prepared from another mother solution in respect to the calibration standards, for stability and preparation checking.

Normally, the laboratories use fixed and inflexible criteria for this control. They define a limit to the percentage difference between the expected and the obtained value, and in low precision techniques they are obliged to increase this value. Assuming the uncertainty associated with the control standard preparation to be negligible when compared to the instrumental uncertainty, the case-to-case interpolation uncertainties can be used as a fit for each case. If the observed confidence interval includes the expected value, there is reason to think that the system is not under control. The instrumental deviation from control can be used as a guide for instrumental checking or as a warning of the inadequacy of the chosen mathematical model for the calibration.

Validation of the uncertainty estimation

Method validation is the process of demonstrating the ability of a method to produce reliable results [8]. An analytical result should be expresed along with a confidence interval and a confidence level. The confidence interval can be described by a mean value and a interval width. Therefore the validation depends on the reliability of the confidence interval width estimation. The accuracy test can be performed exactly only, after that step.

The statistical equivalence between the estimated and the observed values can be used to confirm that quality. The F-test [10] is a good tool for comparing (non-expanded) uncertainties.

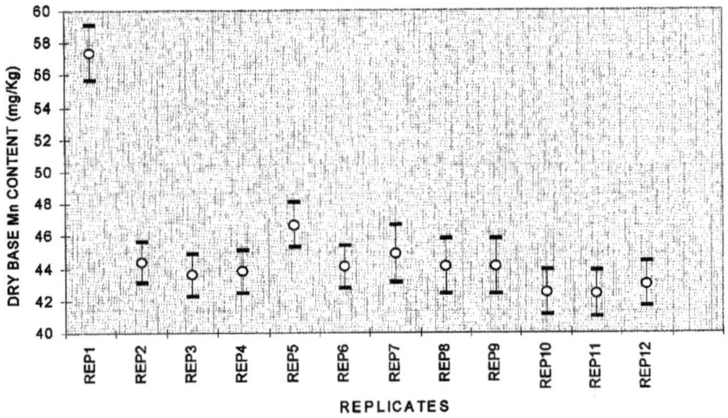

Fig. 2 Repeatability test. The confidence intervals are represented by the average value plus the estimated expanded uncertainty for a 95% confidence level

Application of uncertainty validation schemes

The proposed uncertainty validation method was applied to the dry-base determination of manganese in digested lettuce leaves by electrothermic atomic absorption spectrometry. The proposed quality control scheme was also applied.

The 200-mg samples were digested with nitric acid in a microwave-irradiated closed system [11]. The instrumental determination was performed in a GBC atomic spectrometer with D_2 lamp background correction. A Pd/Mg mixture [12] was used as chemical modifier. The dry-base correction factor was calculated by a parallel assay. The samples were dried in an oven at 60 °C under atmospheric pressure, and the CRM (certified reference material – NIST 1570a) was treated as specified by NIST [13].

Repeatability test

The estimated uncertainties were compared with the experimental ones by an F-test for the 95% confidence level [10]. Figure 2 represents the obtained experimental values associated with the estimated expanded uncertainty (95% confidence level). The coverage factor used was 1.96 for the average effective number of degrees of freedom, df, of 57500. The Eurachem [1] proposal of a coverage factor of 2 is adequate for this case.

The replicates 1 and 5 (REP1, REP5) are consecutive single outliers (Grubbs test) for a 95% confidence level [9]. Therefore, they have not been used for the experimental uncertainty calculation. The two uncertainties are statistically equivalent for the test used (experimental uncertainty: 0.82 mg/Kg for 9 df; estimated uncertainty: 0.73 mg/Kg for 57500 df) at the 95% confidence level.

Accuracy test

The accuracy test was performed with spinach leaves (NIST 1570a) because of their claimed similarity with lettuce leaves in terms of proteins, carbohydrates, fibre and inorganic matter content [14]. The validated uncertainty estimation was used for the comparison of obtained values with certified ones (Fig. 3).

The loss of precision in EAAS with the time of use of furnace is taken into account in the case-to-case interpolation uncertainty calculation. The accuracy is retained with a larger confidence interval.

The overlapping of these intervals indicates that there is no reason to think that our method lacks accuracy. Our analytical method can be considered to perform as badly or as well as the NIST methods. This assumption seems sufficiently valid to consider the method validated.

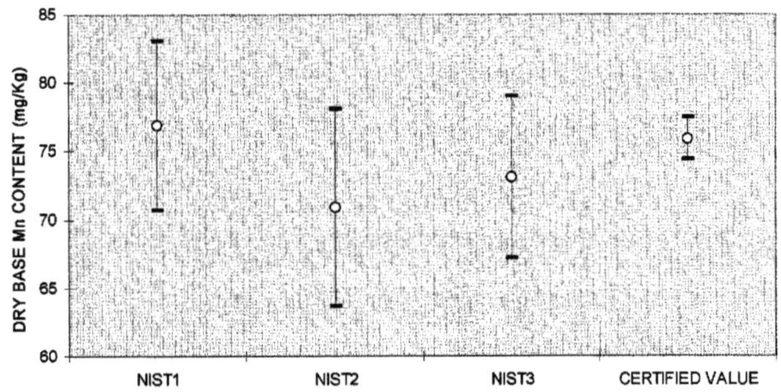

Fig. 3 Accuracy test over spinach leaves NIST CRM. The obtained values (NIST 1, 2 and 3) were associated with a 95% confidence level expanded uncertainty. The certified value is also presented for the same confidence level

Conclusions

The assumption of 100% efficient digestion with negligible uncertainty is valid for the total uncertainty estimation of the presented example. This uncertainty estimation proved to be a valuable criterion for method validation and quality control, which can be tested by a very simple procedure. The easy routine use of an exact treatment in a spreadsheet program can be useful for the more demanding situations. Nevertheless, further approximations can be easily tested.

Acknowledgements Thanks are due to JNICT for financial support to CONTROLAB LDA, for the instrumental facilities that made this work possible, as well as to the teams of INETI and CONTROLAB LDA, for their support.

References

1. Eurachem (1995) Quantifying uncertainty in analytical measurement, version 6
2. ISO (1993) Guide to the expression of uncertainty in measurement, Switzerland
3. Kargten J (1994) Analyse 119:2161–2165
4. Penninckx W, Hartmann C, Massart DL, Smeyers-Verbeke J (1996) J Anal At Spectrom 11:237–246
5. ISO International Standard 8466-1 (1990) Water quality – calibration and evaluation of analytical methods and estimation of performance characteristics – Part 1: Statistical evaluation of performance characteristics, Geneva
6. Analytical Methods Committee (1994) Analyse 119:2363–2366
7. Staats G (1995) Fresenius J Anal Chem 352:413–419
8. Taylor JK (1983) Anal Chem 55:600A–608A
9. Grubbs FE, Beck G (1972) Technometrics 14:847–854
10. Miller JC, Miller JN (1988) Statistics for analytical chemistry (2nd edn). Wiley, UK
11. Deaker M, Maher W (1995) J Anal At Spectrom 10:423–431
12. Soares ME, Bastos ML, Carvalho F, Ferreira M (1995) At Spectrosc 4:149–153
13. NIST (1994) Certificate of Analysis SRM1570a Trace elements in spinach leaves
14. Penninckx W, Smeyers-Verbeke J, Vankeerberghen P, Massart DL (1996) Anal Chem 68:481–489

Accred Qual Assur (1998) 3:189–193

Reports and notes on experiences with quality assurance, validation and accreditation

Robert J. Wells

Validation requirements for chemical methods in quantitative analysis – horses for courses?

Received: 22 September 1997
Accepted: 28 November 1997

R.J. Wells (✉)
Australian Government Analytical Laboratories, 1 Suakin Street, Pymble, NSW 2073, Australia

Abstract Although the validation process necessary to ensure that an analytical method is fit for purpose is universal, the emphasis placed on different aspects of that process will vary according to the end use for which the analytical procedure is designed. It therefore becomes difficult to produce a standard method validation protocol which will be totally applicable to all analytical methods. It is probable that far more than 30% of the methods in routine laboratory use have not been validated to an appropriate level to suit the problem at hand. This situation needs to change and a practical assessment of the degree to which a method requires to be validated is the first step to a reliable and cost effective analytical industry.

Key words Validation · Verification · Veterinary drugs · Anabolic steroids

Introduction and background

All methods used in analytical chemistry are subject to error. Therefore it is vital that each method should be evaluated and tested to ensure that it produces results which make it suitable for the intended purpose. Method validation and verification is the implementation of this evaluation process [1–5]. However, the extent and rigor with which a particular method is evaluated is dependent on the intended use and past experience with the method.

Method validation is the process in which every stage of a new analytical method is subjected to rigorous series of tests to ensure that the method will be able to deliver all the outcomes required of it. The confidence that the method can deliver these outcomes is expressed in terms of statistical probability over the whole analyte concentration range established during the validation process. Verification of a method is a simplified form of the validation process. It involves the testing of a series of method parameters to ensure that a previously validated analytical procedure performs as reported when it is introduced into a new environment where, at the very least, equipment may not be identical to that employed in the initial validation.

Established methods must, as a minimum requirement, be verified when introduced into a laboratory for the first time. Verification, strictly speaking, is also necessary if the method is modified or applied to a new situation, for example a different sample matrix. A new method must be subject to a much more searching series of validation procedures, each one of which adds further confidence in the analytical results obtained. While method validation is mandatory for assurance of analytical quality, the cost to a laboratory is significant. It is therefore important for the financial well-being of a laboratory that validation should adopt no more than those procedures necessary to ensure the analytical quality demanded by a client.

In general, validation processes are universal, but the rigor with which these processes are applied will vary with the intended use of the method. This paper will outline validation principles and discuss the degree to which these principles must be implemented to achieve the necessary confidence in the analytical result obtained.

Types of method, screening vs confirmation

Methods can be classified in a number of ways, but in the present instance an important distinction should be made between screening and confirmatory methods. Confirmatory methods should include some or all of the parameters shown in Table 1. Screening methods require a limited sub-set of all of the parameters used for method validation, but should include those parameters indicated in the first column of Table 1.

Stages of validation

A distinction can be made between establishing the performance characteristics of a method and complete method validation. After the performance of a method has been assessed, complete validation requires evaluation during its application to the intended purpose, using reference methods for comparison, certified reference materials if available, and interlaboratory comparisons in order to obtain a realistic estimate of the uncertainty of routine results. It therefore follows that, where possible, laboratories should not work alone but collaborate in interlaboratory studies.

Important protocols for method validation in the literature have been derived, amongst others, from the Current Good Manufacturing Practice, Code of Federal Regulations, Food and Drug Administration, National Drug Administration, the United States Pharmacopoeia Convention, the American Public Health Association and the International Conference on Harmonization.

The scheme shown in Table 2 is the *ideal* validation procedure for most methods. However, this is a daunting list, which many laboratories would wish to abbreviate without compromising the required analytical quality.

Table 1

Requirements for both screening and confirmatory methods	Additional requirements for confirmatory method
Specificity (selectivity)	Recovery
Sensitivity	Limit of quantitation (LOQ)
Calibration and linearity	Quality control
Accuracy (bias)	Repeatability (between analysts)
Repeatability of method	Reproducibility (between laboratories)
Range	Ruggedness
Limit of detection (LOD)	System suitability test

Table 2

Stage	Process involved
Preliminary stages	
1. Identify requirements	Needs of client
2. Method selection	Literature search, recommendation of colleagues Novel approach, regulatory requirements Staff expertise and requirements for staff training
3. Development candidate method	Preliminary testing
Method validation	
4. Calibration and linearity	Goodness of fit
5. Accuracy (bias)	Comparison with reference method, spiked samples Certified reference materials, collaborative tests
6. Repeatability	Within day, within lab
7. Reproducibility	Day to day, between labs
8. Sensitivity, LOD and LOQ	Instrumental specifications
9. Specificity	Interference studies
10. Recovery	Samples of known concentration, preparation efficiency
11. Quality control	Spiked samples, blanks, critical control points
12. System suitability test	Routine acceptance criteria, control charting
Validity of method to meet required analytical outputs, including calculation of uncertainty for all relevant stages, is now established	
13. Produce method documentation	Write up experimental/validation work
14. Write user instructions	Prepare standard operating procedure (SOP)
15. Management authorisation	
16. Use method	
17. Method review	Updating SOPSs

Metrological requirements

A cornerstone of the development of a method is that all results are traceable. This has been discussed in detail in an earlier paper in this journal by Price [6]. Current requirements of QA and GLP are that all laboratory equipment should have maintenance and calibration records kept in a log book. Such calibration procedures must be traceable to a reference or primary standard. This is true of both general and specialized laboratory equipment, but traceability of some specialized laboratory equipment (e.g. mass spectrometers) may be problematical.

A more serious problem arises when the traceability of the whole analytical process is considered. Complete traceability through to the mole is an excellent ideal but nowhere near achievable at present. Comparability between chemical analyses by use of certified reference materials (CRMs) may be regarded as the initial step in the traceability chain to the mole. Even this relatively small first step is constrained by a number of factors. Some of these are:

- A CRM is usually certified using a specific method. Therefore the same method should be used unchanged for all analyses deemed to be traceable to that CRM.
- There are only a limited number of CRMs available.
- CRMs are often only available for a particular matrix at a particular analyte concentration. Any deviation from either sample matrix or a specific concentration would invalidate traceability.

Added to these constraints are the issues raised by the certainty of a result in terms of both analyte identification and quantitation. The uncertainty of a result is dependent on the analyte concentration. In trace analysis it might be argued that the traceability of a method in identifying a substance would have a traceability chain completely different from that of a method which quantitates the same substance. The traceability chain for a method which both quantitates and identifies a substance could be different again. This differentiation is important in many regulatory analyses in which a zero tolerance for a substance has been set. Under these circumstances, only the presence of the substance has to be established. This problem is discussed in more detail later in this paper.

Practical applications of the validation process

Thus far, this paper has simply summarized information covered in standard texts. Two different practical situations are now described in which it is necessary to emphasize certain parts of the validation process to ensure confidence in the results, while minimizing or even ignoring other processes which have no impact on the analytical outcome. In discussing these separate examples, we hope to demonstrate that a pragmatic rather than a strictly rigid approach to validation must be adopted to economically accommodate various needs.

Establishment and regulation of maximum residue limits (MRLs) for veterinary drugs

One of the two Joint WHO/FAO Expert Committees on Food Additives (JECFA) has the task of recommending appropriate MRLs for veterinary drug residues in edible animal tissues based on toxicological assessments of a particular drug coupled with an accurate knowledge of the depletion of the drug and/or its metabolites from various animal tissues over time. These recommendations then go forward to the Codex Alimentarius for ratification and adoption for regulatory purposes. However, the validation requirements for an acceptable method to generate residue depletion data may be distinctly different from the requirements for a satisfactory regulatory method.

A residue depletion method for a veterinary drug

This is usually developed and used in-house (under GLP) by a company for the specific purpose of measuring residues of the drug in tissues. Seldom is the method tested in another laboratory. Since the drug is administered under controlled conditions, interferences due to the presence of related drugs is never a problem. Indeed, related drugs are often used as surrogates or internal standards in such methods. Therefore only matrix interferences are important in the validation process. Analyte recovery and accuracy of quantification are of prime importance, but since no standard reference materials are available, these parameters must be assessed from spiked samples. The LOQ must be as low as possible and the CV of the method must also be low (ideally $< 10\%$ at the LOQ) and certainly lower than the animal-to-animal variability. The complexity (and therefore the cost) of the method is not a vital factor, nor is the suitability of the method for routine application of great importance. A final correction for recovery is necessary in order to establish the actual residue concentrations in tissue. Often a liquid chromatography (LC) method using pre- or post-column derivatization is the method of choice to meet the analytical objective.

A regulatory method for a veterinary drug

This often has distinctly different requirements. Since residue testing is a factor in world trade, it is vital that there is comparability of results between national laboratories undertaking drug residue monitoring. Thus, an interlaboratory trial of the method is very desirable. Furthermore, it is most unlikely that all regulatory laboratories will be identically equipped, and a regulatory method must be robust and flexible enough to cope with such variation. Another major requirement of the regulator, which is of little concern to the drug manufacturer, is the strong preference for an economical multi-residue method suitable for the quantitation of all drugs of a particular class. An optimum method would combine high efficiency isolation and quantitation with the confirmation of a wide range of drugs of different classes from the same matrix. The selectivity of a method is very important to achieve this aim. The LOQs of each drug included in this ideal method would need to be only 4–10 times lower than the MRL, and the recovery and%CV at the LOQ need not be as stringent as those required in residue depletion studies. The use of related drugs of regulatory interest as internal standards or surrogates would be undesirable in this method, and a gas chromatography-mass spectrometry (GC-MS) or LC-MS method using deuterium-labeled internal standards would probably be the most cost effective way of achieving most or all of the analytical goals.

Clearly satisfactory validation of a residue depletion method would be significantly different in many respects from that demanded by a regulatory method. But who should develop the 'regulatory method' and how is it to be decided if a method is suitable for regulatory purposes?

Codex have requested that, during the evaluation process for recommendation of MRLs, JECFA also ensure that the sponsor of the drug provide a "validated method suitable for regulatory purposes". The drug manufacturer might argue that this is not their responsibility. Codex might reply, with some justification, that if a company seeks registration of a drug then the provision of a method to detect and regulate the usage patterns should be mandatory. The development of a new analytical method, particularly a multi-residue method suitable for all edible tissues, is not an inexpensive enterprise, and, not surprisingly, many companies are reluctant to develop a regulatory method which fully meets the desires of the regulators. This impasse has still to be resolved to the acceptance of each party, with JECFA caught in the middle as both arbiter and judge.

Detection of the use of anabolic steroids in sport

The detection of the illegal use of drugs in sport will be a major priority for the Australian Government Analytical Laboratories (AGAL) for the next 3 years. The analyses carried out by the Australian Sports Drug Testing Laboratory in AGAL (NSW) illustrate a further set of analytical problems which test, to the full, the implementation of a rigid validation protocol. The detection of anabolic steroids in urine will be used as an example.

International Olympic Committee requirements for drug testing are simple. If any banned non-endogenous anabolic agent is unambiguously detected then the sample is positive. Since the protocol employed requires any positive sample to be analysed three times and since the criteria used to declare a sample positive are conservative, false positives should be zero. False negatives are not as easy to control since a decision is made from the initial screen. On top of what is essentially a very sensitive qualitative analysis is a requirement to measure the ratio of testosterone to epitestosterone as a method to detect the illegal use of exogenous testosterone. Again this is not quantitative, but accurate measurement of ratios of analytes at very low levels requires extensive validation information.

Steroids are excreted as the glucuronides and sulfates of the steroid and its metabolites. Enzymic hydrolysis yields the free steroids, which are derivatised and determined by gas chromatography-mass spectrometry (GC-MS). Positive identification of two or three different metabolites is needed to declare a positive. However, only a very few deuterated anabolic steroids or their metabolites are available as pure standards. Indeed, few non-isotopically labeled steroid metabolites are commercially available. Therefore a urine prepared by mixing incurred samples of all the known anabolic agents is used with every batch analysed to monitor the complete analytical procedure. Deuterated testosterone and methyltestosterone are added as surrogate and internal standard respectively. Only the concentration of added substances is known. When the day-to-day or even sample-to-sample variability of GC-MS is recognized together with the fact that all urines are of different complexity, the validation process becomes increasingly difficult.

Deuterated testosterone and DES (diethylstilbestrol) glucuronide allow estimation of the success of the glucuronide deconjugation and the recovery of free steroids. Use of the composite urine ensures that derivative separation and sensitivity are within experimental protocols, and deuterated methyltestosterone tests instrumental performance. Although the LOD of added substances can be measured, those of all the metabolites cannot be accurately obtained. Moreover, in real samples, the LOD for each group of metabolites will be different according to the interfering substances which are present. This is why the criteria used to date to declare a sample positive are so conservative - the athlete always gets the benefit of any doubt. The "traditional" validation of the whole process can only be based on the composite urine used as external standard and is centered around the limit of detection. However, acceptance-rejection criteria based on obtaining satisfactory GC-MS spectra for all designated ions of all target metabolites in the GC-MS run must also be considered as part of the validation process.

The introduction of GC-high resolution mass spectrometry has added another dimension to the problem because much lower signal-to-noise spectra can be obtained and many interferences screened out. This beacons a new era where detection limits will fall and the number of confirmed positives could well rise. Also, will new problems arise in deciding if the method has been adequately validated? That is, is it demonstrably fit for the purpose for which it was designed?

Method validation vs performance criteria validation

The debate about comparability of methods based on performance criteria is still current. The AOAC stand is that results can only be validly compared between laboratories if the same method and preferably the same equipment is used. Although many good arguments can be advanced for this approach, the time and resources required to set up and evaluate interlaboratory tests are often extreme. Also, this approach leads to inertia and reluctance to commit to change or improvement. The obvious argument against performance criteria based validation is the question of what is the bench mark against which the criteria are measured. If results are compared with those obtained for an SRM, then that SRM was probably determined by a standard method. The ideal situation arises when two independent methods give comparable results. This certainly happens – sometimes.

Conclusions

This debate reinforces the need for a rigorous system for traceability in analytical chemistry. The time must come when clients can be confident in the fitness of their results for the purpose for which they were obtained. It has been estimated that at least 30% of all chemical analysis is not fit for purpose. In Australia alone, this probably represents $A 100 000 000 per annum spent on analytical work which is worthless.

However, it is probable that far more than 30% of the methods in routine laboratory use have not been validated to an appropriate level to suit the problem at hand. This situation needs to change and a practical assessment of the degree to which a method requires to be validated is the first step to a reliable and cost-effective analytical industry.

References

1. ISO/IEC Guide 25 (draft 5 1996] General Requirements for the Competence of Testing and Calibration Laboratories. International Organisation for Standards, Geneva
2. Taylor JK (1993) Anal Chem 55:600A–608A
3. Green JM (1996) Anal Chem 68:305A–309A
4. Thomson M, Wood R (1995) Pure Appl Chem 87:649–666
5. Guideline for collaborative study of procedure to validate characteristics of a method of analysis (1989) J Assoc Off Anal Chem 72:694–704
6. Price G (1996) Accred Qual Assur 1:57–66

Accred Qual Assur (1998) 3:194–196

Reports and notes on experiences with quality assurance, validation and accreditation

Raluca-Ioana Stefan
Hassan Y. Aboul-Enein

Validation criteria for developing ion-selective membrane electrodes for analysis of pharmaceuticals

Received: 18 September 1997
Accepted: 17 November 1997

R.-I. Stefan
Department of Analytical Chemistry, Faculty of Chemistry, University of Bucharest, Blvd. Republicii No. 13, 70346, Bucharest-3, Romania

H. Y. Aboul-Enein (✉)
Bioanalytical and Drug Development Laboratory, Biological and Medical Research (MBC-03), King Faisal Specialist Hospital and Research Centre, P.O. Box 3354, Riyadh 11211, Saudi Arabia
Fax: +966-1-442 7858
e-mail: enein@kfshrc.edu.sa

Abstract The problem of validation criteria for developing ion-selective membrane electrodes for the analysis of pharmaceuticals arises from the connection between the reliability of ion-selective membrane electrodes construction and the reliability of the analytical information. Liquid membrane selective electrodes are more suitable for validation than the solid variety. The influence of the stability of ion pair complexes from the membrane on various parameters (e.g. response, limit of detection, and selectivity) is discussed. Validation criteria are proposed.

Key words Ion-selective membrane electrodes · Reliability of construction · Reliability of analytical information · Validation criteria · Analysis of pharmaceuticals

Introduction

For pharmaceutical analyses, it is necessary to have reliable methods. Results obtained using ion-selective membrane electrodes are the best because of the simplicity, rapidity, and accuracy of direct and continuous measurement of the activity of the ions in the solution. Another very important reason for the selection of electrochemical sensors for pharmaceutical assay is non-interference of by-products when the purity of a raw material is to be determined.

Taking into account the definition of the validation of an analytical method given in the US Pharmacopoeia: "Validation of an analytical method is the process by which it is established, by laboratory studies, that the performance characteristics of the method meet the requirements for the intended analytical applications. Performance characteristics are expressed in terms of analytical parameters."[1]. We can conclude from this that the performance of the method can be evaluated through the analytical parameters of the chemical sensors and by recovery studies.

Using electrochemical sensors and considering the sources of uncertainty given by Pan [2], viz. homogeneity, recovery, analysis blank, measurement standard, calibration, matrix effect and interferences, measuring instrument, and data processing, the main uncertainty sources i.e., the homogeneity and the matrix effect, are eliminated.

The validation criteria for ion-selective membrane electrodes (ISMEs) can be established only by taking into account their reliability and its effect on the reliability of the analytical information that it is obtained.

The reliability of the construction of ISMEs

For the construction of ISMEs, solid membranes and liquid membranes are proposed. The construction of solid membrane ISEs is based on the insertion of electroactive material (ion pair complex), or only of a counter ion into a polyvinyl chloride (PVC) matrix. The PVC type as well as the plasticizer used for the membrane construction affect the reliability of the electrode's response. Thomas proposed some rules for solid membranes based on a PVC matrix construction [3--5]. The procedure most often used for solid membrane construction is known as the Moody and Thomas procedure [5], as this assures the best reliability of solid membrane electrodes.

Reliable liquid membrane electrodes are obtained by impregnating a graphite rod of spectral purity with the solution that contains the ion pair complex dissolved in the appropriate solvent [6]. The rod is attached to the end of a polytrifluoroethene (PTFE) tube.

Why liquid membrane ion-selective electrodes are more suitable for validation than solid membrane ion-selective electrodes

The first principle for ISME validation is the reproducibility of their construction. The construction of the solid membrane is not reproducible because the electroactive material is not uniformly distributed in the membrane. On the other hand, the impregnation process in liquid membrane construction causes the electroactive material to be uniformly distributed in the graphite rod surface because of the uniformity of distribution of the pores in the graphite.

The second criterion for non-selective electrodes is the simplicity of the membrane construction. For a solid membrane, smaller quantities of electroactive materials, plasticizers, and solvent (usually tetrahydrofuran) are used in a large quantity of PVC. Homogenization between them is not easy to achieve. However, after homogenization, it is necessary to satisfy many strict conditions to obtain a solid membrane with a given uniform thickness. This is very hard to achieve, and accordingly it affects the electrode response. The impregnation of the graphite rod with the ion pair complex solution is done by immersion of the graphite rod in the solution, and thus the thickness of the impregnation zone is constant for the same ion pair complex every time.

The third criterion for ISEM construction is the rapidity of the construction process.

For solid membrane construction one must wait more than 24 h to obtain the membrane and another 24 h or more immersion in the inner solution in order for it to be suitable for characterization, Whereas ISEs with liquid membrane can be obtained and characterized in the same day, because 1 h is required for the extraction of the ion pair complex solution and 12 h for the impregnation of graphite rod with the solution.

Taking into account the above-mentioned criteria, one must recognize that ISEs based on liquid membranes are more suitable for validation, because their construction is reliable, simple, and rapid.

The connection between the reliability of developed ISMEs and the reliability of obtained analytical information

The good relationship between the reliability of developed ion-selective membrane electrodes and the reliability of obtained analytical information assures the validation of the ISMEs for analysis.

To obtain accurate results it is necessary to use reference materials [7] and adequate procedures for the calibration of an ISME [8]. The response of the ISME is connected with the stability of the ion pair complex, because the stability of the complex is affected by the solvent used for ion pair complex extraction. In PVC matrix membranes, the PVC and plasticizer mixture plays the same role as the solvent for the ion pair complex, because it affects the stability of this complex. From this point of view, the PVC matrix membrane assures a response value more than 55 mV/decade. The response value of the liquid membrane electrode is generally less than the Nernstian one. Usually it shows responses in the 50–55 mV/decade range. But a response having the 50 mV/decade value assures good accuracy and also the reliability of the analytical information.

The limit of detection is also related to the response of the ISME:

$$\mathrm{p}X_1 = \frac{E^\circ}{S}$$

$\mathrm{p}X_1 = -\log[\mathrm{X}_1]$ where $[\mathrm{X}_1]$ is the limit of detection, E° is the standard potential, and S is the slope. It follows that a good limit of detection of 10^{-6} to 10^{-10} mol/L can be assured by a high E° value and a low S value. If we consider the minimum value for S, 50 mV/decade, the minimum E° value for a slope of 50 mV/decade must be 300 mV to obtain a 10^{-6} mol/L for the limit of detection. The E° value is given by the free energy for the ion pair complex formation, and it also depends on the stability constant of the ion pair complex.

The best limit of detection is assured by the ion-selective electrodes with a liquid membrane. The lower limits of detection are necessary for pharmaceutical analysis, especially for in vitro and in vivo dissolution tests, when low quantities of ions must be detected. The working range is also a very important parameter for ISMEs. It must be not less than two concentration decades.

Selectivity is one of the essential characteristics of the membrane electrodes. Accordingly, it is important that by-products, degradation products, metabolites, and compressing components (excipients) do not interfere, and thus the ISMEs can be successfully used for raw material assays as well as for raw material determinations in pharmaceutical formulations. To improve the selectivity, it is necessary to choose another counter ion for membrane construction, because the selectivity of membrane electrodes can be explained through the stability constants of the ion pair complexes between the counter ion and the ions from the solution. To be selective over one ion from the solution it is necessary for the stability constant of the ion pair complex to be less than the stability constant of the ion pair complex obtained between the main ion and counter ion.

The response time is one of the main factors for ISME validation, because these electrodes must be used to perform dissolution tests of pharmaceuticals and potentiometrical titrations. Recovery studies are a very important means of documenting the reliability of a method. Recovery studies give information about accuracy and repeatability of the analytical method. For accurate analytical information, both the repeatability and the reliability are given by relative standard deviation (RSD) values. When ISMEs are used for pharmaceuticals analysis, the reliability must have the maximum value and RSD $\leq$1.5%.

Validation criteria of ISME

An ISME can be validated for pharmaceuticals analysis only if:

1. the best counter ion (giving the ion pair complex with the best stability) is chosen,
2. the best matrix (PVC, liquid-solvent) for the electroactive material is used,
3. the biocompatibility of materials is correlated with the response characteristics of the ISME,
4. the counter ion assures a slope of 50 mV/decade (minimum value), a limit of detection of 10^{-6} mol/L (minimum value), a large working range, a response time less than 1 min, and the maximum selectivity in preference to the by-products, degradation products, compression compounds, and metabolites.

Conclusions

The validation of ion-selective membrane electrodes is based on the reproducibility of their development, the simplicity and rapidity of their construction, and their response characteristics. Typically, these response characteristics include: minimum 50 mV/decade for the slope, 10^{-6} mol/L for limit of detection, large working range, and low response time, which can be assured only by the best counter ion and matrix (PVC and plasticizer for solid membrane electrodes and solvent for liquid membrane electrodes).

An RSD of more than 1.5% for a recovery test cannot be permitted. RSD values assure both the accuracy and the reliability of the analytical information.

References

1. US Pharmacopoeia, USPXXIII (1994), <1225> Validation of Compendial Methods, USP-Commission, United States Pharmacopeial Convention, Inc., Rockville, MD, USA, p 1982
2. Pan XR (1996) Accred Qual Assur 1:181–185
3. Thomas JDR (1994) Analyst 119:203–208
4. Ghauri MS, Thomas JDR (1994) Anal Proc Including Anal Comm 31:181–183
5. Craggs A, Moody GJ, Thomas JDR (1974) J Chem Educ 51:541–544
6. Cosofret VV (1978) Rev Roum Chim 23:1489–1493
7. Aboul-Enein HY (1996) Accred Qual Assur 1:176
8. Buck RP, Cosofret VV (1993) Pure Appl Chem 65:1849–1858

Accred Qual Assur (1998) 3:412–415

Reports and notes on experiences with quality assurance, validation and accreditation

Stephan Küppers

Is the estimation of measurement uncertainty a viable alternative to validation?

Received: 30 April 1998
Accepted: 3 July 1998

Presented at: Analytica Conference 98, Symposium 2: "Uncertainty budgets in chemical measurements", Munich, 21–24 April 1998

S. Küppers (✉)
Schering AG,
In-Process-Control,
Müllerstrasse 170–178,
D-13342 Berlin, Germany
Tel.: +49-30-468-1-7819
Fax: +49-30-468-9-7819
e-mail: stephan.kueppers@schering.de

Abstract Two examples of the use of measurement uncertainty in a development environment are presented and compared to the use of validation. It is concluded that measurement uncertainty is a good alternative to validation for chemical processes in the development stage. Some advantages of measurement uncertainty are described. The major advantages are that the estimations of measurement uncertainty are very efficient, and can be performed before analysis of the samples. The results of measurement uncertainty influence the type of analysis employed in the development process, and the measurement design can be adjusted to the need of the process.

Key words Chemical analysis · Development process · Measurement uncertainty · Validation · Practical examples

Introduction

The "know how" in quality assessment has grown over the last few years. Quality management systems are improving and validation procedures are being optimized. But the problem that validation procedures are time consuming still remains. Often validations are performed too late. Especially in a development environment where the advance from one development step to the next is based on a decision resulting from the previous step. Therefore, valid analytical data is needed directly after each method development.

The estimation of measurement uncertainty is an alternative to validation if a quantitative result is used for the assessment of the development process. In this case only one parameter of the validation is needed. The precision of the analysis is sometimes taken as a quality figure instead of a validation. But precision cannot replace validation because in this case the environment, i.e. the sample preparation is neglected and precision is only a representative of one contribution to the uncertainty of measurement: in high performance liquid chromatography (HPLC) usually the uncertainty contribution of the sampler.

To illustrate the point, two examples taken from a process research environment in a pharmaceutical company are presented. It is demonstrated that measurement uncertainty has practical advantages compared to validation.

Examples of process development

The first example presents a typical process from our research environment, where a chemical synthesis is transferred from development to production. In this situation validation of the chemical process is performed throughout, including variations of process parameters for intermediates that are not isolated. For these intermediates a pre-selected analytical method is normally

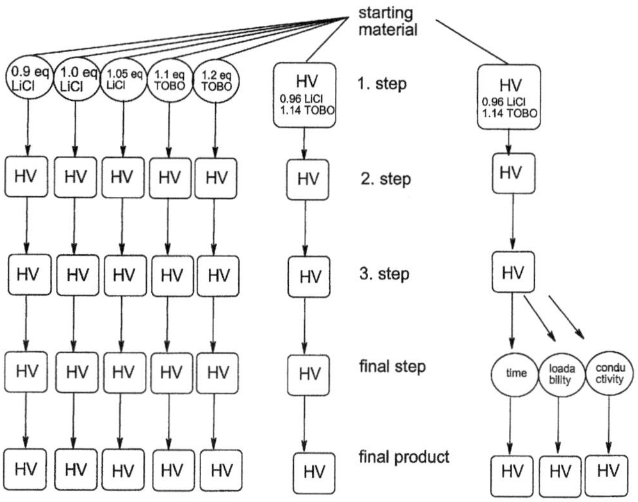

Fig. 1 The synthesis and variations performed in a process research validation shown schematically. HV stands for performed according to the manufacturing formula. The example shown illustrates the first step of the chemical synthesis. *Circles* show the variations of the two parameters tested in the validation (LiCl and TOBO are internal names for chemicals used in the synthesis, eq stands for equivalent)

available, usually a method that has been employed for some time. Because no validation report is requested by regulatory authorities, no formal validation is performed.

An example of a typical validation scheme for a chemical synthesis in process research is given in Fig. 1. In this example the process research chemist sets up a number of experiments. The most important question presented by the chemist to the analytical department is: Is the variability of the analytical process small compared to the variation performed in the process validation [1, 2].

The original manufacturing formula (HV) and five variations are performed in the first step of the synthesis. Six samples are analysed. The results of these six analyses are used to assess the validation of this process step. In this case validation of the analytical method is a prerequisite for any decision that is made about the validity of the process. This information is needed before the process research chemist can start variations of the process otherwise it is possible that the data received cannot be assessed. The difficulty of assessing the data of the process validation results from the fact that the data is influenced by the analytical method and the uncertainty of the chemical process. If the uncertainty of the analytical method is larger or in the same range as the variations of the chemical process, assessment of the data is not possible.

To assess uncertainty, a type B estimation of uncertainty is performed. After analysis of the samples the type B estimation can be verified by a type A estimation analogous to the concept presented by Henrion et al. [3]. In our laboratory, a type B estimation can be performed with confidence because we have performed this type of analysis (an HPLC method with external standard calibration) about 50 times before. A control sample is included in every analysis and the results are plotted on a control chart. The control chart, representing the total measurement uncertainty for the analytical method, is then reviewed for the estimation of uncertainty. The standard deviation for the assay of a control sample for 50 analyses was 1.51%. An example of the way in which a control sample can be used for measurement uncertainty is presented in detail in [1]. The typical uncertainty components for this type of analysis are (Table 1):

— inhomogeneity of the sample and sampling
— weighing of reference materials (about 50 mg)
— weighing of samples (about 50 mg)
— uncertainty of the instrumentation (sampler uncertainty)
— evaluation uncertainty (uncertainty caused by integration)
— uncertainty of reference materials.

Table 1 First step in the estimation of the measurement uncertainty of high performance liquid chromatography (HPLC) analysis as a method used for the assessment of a chemical process

Uncertainty component	Min	Max
Inhomogeneity of the sample and inhomogeneity of sampling	<0.2%	0.5%
Weighing of the reference materials or the sample (about 50 mg)	<0.1%	0.1%
Uncertainty of the instrumentation (sampler uncertainty) and reference material(s) depending on the number of injections	0.5%	1.5%
Evaluation uncertainty (uncertainty caused by integration)	<0.1%	0.1%
Uncertainty of the reference material(s)	–	–

The case presented here is simple because all six samples can be analysed as one set of samples in one analytical run. For a single run the estimation of uncertainty can be reduced to the calibration of the method with an in-house reference material and the uncertainty of the analysis itself. In this example inhomogeneity of the sample, the evaluation uncertainty and the uncertainty of the reference material (i.e. inhomogeneity) as given in Table 1 can be neglected because the results of only one analysis are compared. The weighing uncertainties can be neglected because they are small compared to the sampler uncertainty.

If the measurement uncertainty is estimated using the parameters in Table 1 but for a longer time period

Table 2 Results of the estimation of the measurement uncertainty for HPLC analysis

	Option 1[a]	Option 2[b]
Calibration	1%	1%
Analysis of samples	1.5%	1%
Total uncertainty (received from uncertainty propagation)	1.8%	1.4%

[a] Option 1: two weights of the calibration standard with six injections for each of them, two weights of the sample with two injections for each weight
[b] Option 2: two weights of the sample with four injections for each weight

Table 3 Comparison of the results of the estimation and the analysis (type B estimation compared to type A estimation)

	Estimation	Found (one example)
Calibration	1%	0.8%
Analysis of samples	1%	0.8%
Total uncertainty (by uncertainty propagation)	1.4%	1.14%

as for example covered in the control chart, the inhomogeneity of the sample has to be included in the estimation. Using the mean values between the min and max column in Table 1 the uncertainty estimated by uncertainty propagation is 1.46%, which is close to the value found in the control chart. In the case presented here the estimation of measurement uncertainty can be performed in only two steps as shown in Table 2. Calibration and analysis of samples represent the major uncertainties and combined they provide the complete uncertainty of our experiment; in this case the uncertainty of the HPLC sampler. The influence of the HPLC sampler is known, therefore, there are two options to perform the analysis.

Together with the customer it was decided to perform the analysis according to option 2. The samples were analysed in one analytical run. The result is shown in Table 3. The result of the estimation compares well with the "found" results. The found uncertainty is smaller than the estimation. Assessment of the results of the validation of the manufacturing formula becomes easier from the customers point of view because the customer is able to deceide if a variation of his result is related to his process or to the uncertainty of the analytical method. Additionally, the influence of the individual contributions to uncertainty becomes smaller because of the uncertainty propagation. Therefore, the difference between the estimated and found uncertainty becomes smaller with an increasing number of parameters that influence uncertainty.

The estimation of uncertainty replaces a full validation of the analytical method. It generates the necessary information at the right time. The statistical information received from the analysis can be used for the interpretation of the data and finally the analysis is designed to the customers needs. In this case measurement uncertainty is a good alternative to validation.

The second example illustrates the determination of water content which is an important characteristic for chemical substances and is needed in many chemical reactions. It is usually determined by Karl-Fischers (KF) titration. The water content determined in our laboratory ranges from <0.1% to about 30%. It is widely known that KF water titration's may be influenced by the sample and, depending on the range, some other parameters may significantly affect the uncertainty.

Because the concentration of a reagent is often determined on the basis of the water content of the reaction mixture, uncertainty information for the water determination is needed. The problem in a development environment is that various synthesis routes are tested. If salts are used in a chemical reaction it is usual that chemists test different counter ions for the optimization of the synthesis. However, different salts are rarely tested in the analytical department. One of the problems of the variation of counter ions is that the hygroscopicity of the salts is often different.

Two independent steps have to be followed:

1. Substance dependent influences have to observed. In most cases the chemical structure of the component is known and therefore serious mistakes can be avoided. Titration software and various KF reagents have to be available and standard operation procedures have to be established.
2. The individual uncertainty has to be considered. Therefore a preliminary specification has to be set and a type B estimation of uncertainty can be used to show if there is any problem arising from the data and the specification limit [4].

The first step is a general task using the appropriate equipment and has been established in our laboratory. However, the second part needs to be discussed in detail.

Suppose there is a chemical reaction with reagent A where:

$$A + B \rightarrow C. \quad (1)$$

The alternative reaction for A may also be with water to D.

$$A + H_2O \rightarrow D. \quad (2)$$

The reaction with water is often faster than with B. Because of the low molecular weight of water 0.5% (w/w) of water in B may be 10 mol %. Therefore an excess of at least 10 mol % of A might be needed to complete the reaction. The water content is determined in the

Table 4 Estimation of the measurement uncertainty for the titration of water (example performed manually)

Results from the titrations (% (w/w) from the weight of the sample)	0.20%; 0.23%	0.215% +0.0150%
Minimum uncertainty of 1.1%	1.1%	+0.00236%
Hygroscopicity (estimated on the basis of experience with amin compounds)	5%	+0.00621%
Uncertainty of the titer	1%	+0.00124%
Reaction of the sample with the solvent	5%	+0.00621%
Result reported including uncertainty		0.25%

analytical department. For example the value determined by KF titration is 0.22% (w/w) from two measurements with 0.2% (w/w) and 0.23% (w/w) as the individual values. The question that has to be asked is: Is 0.22% of the weight always smaller than 0.25% (w/w)? What we need is the measurement uncertainty added to the "found" value. If this value is smaller than the limit (0.25%) the pre-calculated amount of the reagents can be used.

The model set up consists of two terms:

$$Y = X^* (1 + U)$$

where X is the mean value of the measurements plus the standard uncertainty. U is the sum of the various influence parameters on measurement uncertainty:
- hygroscopicity
- uncertainty of the titer
- reaction of the KF solvent with the sample
- a minimum uncertainty constant of 1.1% (taken from the control chart of the standard reference material) covering balance uncertainties, the influence of water in the atmosphere and the instrument uncertainty with the detection of the end of titration.

The factors for the three contributions mentioned above are estimated on the basis of experience. The calculation is performed using a computer program [5]. This makes the decision easy and fast. In this case the type A estimation on the basis of the results and the type B estimation of the influence factors are combined. An example is given in Table 4 in a compressed form.

The alternative would be an experimental validation. In this case the uncertainty estimation has proven to be a very useful alternative to validation, although, on the basis of experience, the estimate of hygroscopicity is difficult and may lead to incorrect values.

Conclusions

Method validation is a process used to confirm that an analytical procedure employed for a specific test is suitable for the intended use. The examples above show that the estimation of measurement uncertainty is a viable alternative to validation. The estimation of measurement uncertainty can be used to confirm that an analytical procedure is suitable for the intended use. If the estimation of measurement uncertainty is used together with validations both the uncertainty estimation and the validation have their own place in a developmental environment. The major advantages of measurement uncertainty are that it is fast and efficient. Normally, if the analytical method is understood by the laboratory very similar results are found for the estimation of uncertainty and for the classical variation of critical parameters, namely, validation. The decision on how to perform a validation should be made on a case to case basis depending on experience.

Acknowledgements Fruitful scientific discussions with Dr. P. Blaszkiewicz, Schering AG, Berlin and Dr. W. Hässelbarth, BAM, Berlin are gratefully acknowledged.

References

1. Küppers S (1997) Accred Qual Assur 2:30–35
2. Küppers S (1997) Accred Qual Assur 2:338–341
3. Henrion A, Dube G, Richter W (1997) Fres J Anal Chem 358:506–508
4. Renger B (1997) Pharm Tech Europe 9:36–44
5. Evaluation of uncertainty (1997) R. Metrodata GmbH, Grenzach-Whylen, Germany

Accred Qual Assur (1999) 4:473–476

Reports and notes on experiences with quality assurance, validation and accreditation

R.C. Díaz
M.A. Sarasa
C. Ríos
J.J. Cuello

Validation of an HPLC method for the determination of p-dichlorobenzene and naphthalene in mothrepellents

Received: 21 December 1998
Accepted: 4 May 1999

R.C. Díaz (✉) · M.A. Sarasa · C. Rios · J.J. Cuello
University of Zaragoza, Analytical Chemistry Department. Crtra. Zaragoza, s/n, 22071 Huesca, Spain
e-mail: cdiaz@posta.unizar.es
Tel.: +34-9-74 239320
Fax: +34-9-74 239302

Abstract The determination of dichlorobenzene and naphthalene in commercial repellents used in Spain has been validated. This was done using an isocratic regime, to test the reverse -phase HPLC system with acetonitrile: water 65:35 (v: v) as the mobile phase, at 20 °C. This technique is proposed for the modular validation of the HPLC system . The results obtained with this method show good agreement with the results provided by the manufacturers of the mothrepellents.

Key words Validation · HPLC · p-Dichlorobenzene · Naphthalene · Mothrepellents.

Introduction

The most frequently used commercial mothrepellents in Spain contain p-diclorobenzene (p-DB) or naphthalene (N) as the active component (98–99% w/w), the rest being perfume, approximately 1–2%.These values are generally determined by gas cromatography [1–5]. We propose an alternative analytical method, based on determination of the active components by HPLC in reverse phase, using an isocratic regime, with diphenyl (DP) as an internal standard. The results obtained show good conformity with the certified values of the active components supplied by the manufacturers of the mothrepellents. The modular approximation [6] is proposed for the validation of the method, that includes the validation of each part of the system (pump, injector, etc.).

HPLC validation systems

The stages involved in the validation of the HPLC system, in chronological order, are as follows [7–10]:

Column

The column used for the validation was a reverse phase Nucleosil 120 C-18. Methanol was used as the mobile phase, to achieve a resolution of Rs >1.5. The mobile phase flow rate was 1 ml/min, the injection volume was 10 μl, with detection at 250 nm. The determination was accomplished at ambient laboratory temperature (20 °C). The trial was repeated 10 times. The mixture to determine the real flow consisted of nitromethane, anthracene, pirene and perilene, dissolved in methanol. We recommend that the proportions are adjusted so that each peak corresponds to approximately 25% of the area with respect to normalized areas. As an illustration, for 0.5 units of absorbance (AU), the cited solution will contain: nitromethane (1000 μg ml^{-1}), anthracene (2 μg ml^{-1}), pirene (33 μg ml^{-1}) and perilene

(12 μg ml^{-1}). Uracil (5 μg ml^{-1}) was also added as a not retained reference component.

Pump

Flow accuracy

The flow volume exiting of the detector was measured using water of chromatographic quality. The eluent volume was measured over a 10 min period, in a graduated manometer and for different flow rates (0.5, 1.0 and 2.0 ml min^{-1}) the time elapsed was measured with a precision chronometer. The flow rate was then calculated. This test was repeated every 10 days to evaluate the reproducibility of the flow. An accuracy of up to 3% of the theoretical value was recorded. The observed variations should be to the critical parts of the pump (fundamentally the valves).

Calculation of the short-term flow

As of the obtained chromatogram obtained from the sample test, is calculated, for each one of its components, the variation coefficient of the values of the normalized areas, for ten injections. The expression for the global variation coefficient of the mixture is shown in Table 1. The tolerance turned out to be inferior to 1.5% which is the tolerance recommended in the literature.

Measurement of the flow during a long run

The variation coefficient of the retention times, ($t'_R = t_R - t_{uracil}$) corrected, for ten injections was calculated from the chromatogram obtained for the sample test. The global variation coefficient turned out to be inferior to 1%. The equation used for the calculation appears in Table 1.

Injector

All determinations were carried out by manual injection.

Reproducibility

The reproducibility was calculated for each component of the test mixture by sixtuplicated injections of 20 μl. The global variation coefficient was calculated usingthe equation shown in Table 1.

Linearity

A calibration curve was constructed using a solution of anthracene in methanol (2 μg ml^{-1}), by quintuplicate injections of 10, 20 and 30 μl. The correlation coefficient was 0.9890 and the precision for each group of injections was 2.0%.

UV Detector

Accuracy of the wavelength

The cell of the detector was filled with a chromophore solution of uracil in methanol (15 μg ml^{-1}) A sweep was performed between 250 and 265 nm. The wavelength corresponding to the maximum absorbance must be 257±2 nm.

Linearity

In order to test the linear response range of the detector, a standard solution of propylparaben in methanol (25 μg ml^{-1}) was prepared, that represents 2 AU. A calibration curve was constructed using 5, 10, 15, 20 and 25 μg ml^{-1}, beginning with the most dilute solution.

Quantification limit

A standard solution of anthracene in methanol (2 μg ml^{-1}) was used. The same methanol used for the mobile phase and solvent of the test solution was used a a blank control. With the integrator set at minimum attenuation (maximum sensibility), the noise was determined for six injections of methanol (10 μl), calculating s_b. The value 10 s_b gave a signal level related to the quantification limit, obtained from the standard solution of anthracene, with the necessary dilution to ob-

Table 1 Variation coefficients (in %) obtained for the determination of flow rate and reproducibility of injections

Parameter to determine	Equation	Tolerance
Short-term flow	Global CV = ((CVi)norm.area/4	<1.5
Flow during a long run	Global $CV = (CV_i)t'_{R}/4$	<1
Reproducibility of injection	Global $CV = [(CV_{abs.area})^2 - (CV_{norm.area})^2]/4$	<1.5

tain a final concentration to satisfy the requirement 10 s_b.

Data processing

Calibration of the software

The set of chromatograms corresponding to the four cited standards was processed and compared to the obtained results and weighted according to the precalculations. There were the differences between the two calibration curves [slope, interception at the origin, R (correlation coefficient), R^2 and standard deviation].

Linearity

A correlation exists between the response of the convertor of the analoque signal and the exit of the detector. It is a registered response of the data acquisition system, for each one of the accomplished readings of the standards of absorbance, analysing the correlation between the values obtained with the system from data process and the accomplished readings directly in the detector.

Uncertainty calculation

Components of type A

They are associated with the calibration curve, s_c. The equation of the straight line obtained is: $y=b+mx$, where y is the area of the peak of each standard in the chromatogram (expressed in area units), b is the interception at the origin and x the concentration of the standard (expressed in μg ml^{-1}). $s_m=s_r/s^{1/2}$ is the corresponding deviation from the slope, s_r the standard deviation of the differences between the obtained value and the real value and s the corresponding to concentrations.

Components of type B

The contribution to the total uncertainty associated with the repetitions of the certified standard was determined by s_a. N is the number of samples of Certified Reference Materials (CMRS) and N the number of repetitions for each sample. Also they contribute to the number of repetitions n of the number N of samples CMRS, with the term s_a. There exists a inaccuracy intrinsically related to the HPLC equipment used. u_e is the contribution to the total uncertainty. Also associated with the certified standards, of agreement to the calibration certificate, that is represented by s_{sc} (I_o is the uncertainty interval and k the uncertainty factor, in this case 2). u_d is related to the successive dilutions effected to prepare the corresponding standards for construction of the calibration curve, with s_d the standard deviation associated with standard most diluted, since is greater the one which contribution adds to the uncertainty. The expressions of the corresponding uncertainties appear in Table 2.

The global variance is calculated from: $s_y^2=[(s_a^2+u_e^2+u_d^2+u_{sc}^2)/m^2]+s_c^2/m^2+s_m^2\,(y-b)^2/m^4$.

The total uncertainty is calculated by the equation $I=ks_y^2$, where $k=2$ (with a confidence level of the 95%).

Table 2 Expressions of the various contributions to total uncertainty

Contribution to uncertainty	Equation foruncertainty
Repetitions of the certified standard	$s_a^2=s^2(1/N+1/n)$
Calibration curve	$s_c^2=s_r^2[1/N-(\Sigma x_i)^2/(\Sigma x_i^2)]$
Precision of the HPLC equipment	$u_e^2=s_e^2/3$
Precision of the CMRS	$u_{sc}^2=(I_o/k)^2$
Dilution of the reference standard	$u_d^2=s_d^2/3$

Table 3 Chromatography and integrator conditions

Column: Nucleosil 120 C-18, 25 cm × 0.46 cm × 5 μm of film thickness
Mobile phase: acetonitrile/water; 65:35 (v/v)
Mobile phase flow: 1.2 ml min^{-1}
Temperature: 20 °C
Injection: 10 μl
Attenuation: 7
Threshold: 5

Experimental procedure

Standard solutions p-DB or N (3200, 5300, 7400 and 9300 μg ml^{-1}) were prepared in 25-ml flasks. DP (2 ml) was added to flask as an internal standard, and the final volume was made up to 25 ml with acetonitrile. The test samples were prepared with 3500 μg ml^{-1} of the corresponding mothrepellent, with 2 ml of DP and made up to 25 ml with acetonitrile.

The optimized chromatographic and integrator conditions appear in Table 3.

Table 4 Results obtained (expressed in % w/w) in the determination of p - dicloroben-zene (p-DB) or naphthalene (N) in test and commercial samples of mothrepellents

Commercial name	Certified (p-DB)	Certified (N)	Result
Polil (a)	98–99	–	98.35 ± 0.11
p-DB (b)	>99	–	99.90 ± 0.06
Naftalina (c)	–	>99	99.86 ± 0.04
Camfolina (d)	–	97–98	97.36 ± 0.14

(a) Cruz Verde – Legrain
(b) Bayer AG
(c) Industrial Química del Nalón
(d) Vanossa

Results and discussion

The standard components were identified by their retention times (8.09 min for N, 9.63 min for p-DB and 11,28 min for DP), from their corresponding calibration graphs. The equation used for the determination of p-DB was: Area of the peak p-DB/Area of the peak DP = b + m [p-DB]/[DP], and Area of the peak N/Area of the peak DP = b + m [N]/[DP] was used for the determination of N. By interpolating the values obtained for the mothrepellent samples, the content of p-DB or N in the commercial products (Table 4). The values obtained can be compared with the values supplied by the manufacturers (for five repetitions of each sample) and are shown in Table 4.

Conclusions

A simple and fast HP method for the determination of p-DB and N in mothrepellents manufactured in Spain has been validated. This was achieved by modular validation of the HPLC system. The results obtained show good agreement with the values provided by the manufacturers.

References

1. Heinen H (1988) Analysis method of dichlorobenzenes. Bayer AG, Leverkusen, Germany
2. Gaset P (1994) Analysis method of p-dichlorobenzene. Cruz Verde – Legrain, Barcelona, Spain
3. NIOSH (1978) Manual of Analytical Methods. Vol. 4
4. García E (1987) Analysis methods for naphthalene. Industrial Química del Nalón, Oviedo, Spain
5. Ortiz de Urbina G (1993) Analysis of refined naphthalene. INASMET, San Sebastián, Spain
6. Tetzlaff RF (1992) Pharm Technol 16:70 – 83
7. Waters (1993) Guide for the system qualification of HPLC automated with detection UV
8. Bedson P, Rudd D (1999) Accred Qual Assur 4 : 50 – 62
9. EURACHEM (1995) Quantifying uncertainly in analytical measurement. Teddington, UK
10. Huber L (1998) BioPharm 11:41, 65–66

Accred Qual Assur (2000) 5:47–53

Vicki J. Barwick
Stephen L.R. Ellison

The evaluation of measurement uncertainty from method validation studies

Part 1: Description of a laboratory protocol

Received: 10 April 1999
Accepted: 24 September 1999

V.J. Barwick (✉) · S.L.R. Ellison
Laboratory of the Government Chemist,
Queens Road, Teddington, Middlesex,
TW11 0LY, UK
e-mail: vjb@lgc.co.uk
Tel.: +44-20-89437421
Fax: +44-20-89432767

Abstract A protocol has been developed illustrating the link between validation experiments, such as precision, trueness and ruggedness testing, and measurement uncertainty evaluation. By planning validation experiments with uncertainty estimation in mind, uncertainty budgets can be obtained from validation data with little additional effort. The main stages in the uncertainty estimation process are described, and the use of trueness and ruggedness studies is discussed in detail. The practical application of the protocol will be illustrated in Part 2, with reference to a method for the determination of three markers (CI solvent red 24, quinizarin and CI solvent yellow 124) in fuel oil samples.

Key words Measurement uncertainty · Method validation · Precision · Trueness · Ruggedness testing

Introduction

In recent years, the subject of the evaluation of measurement uncertainty in analytical chemistry has generated a significant level of interest and discussion. It is generally acknowledged that the fitness for purpose of an analytical result cannot be assessed without some estimate of the measurement uncertainty to compare with the confidence required. The *Guide to the Expression of Uncertainty in Measurement* (GUM) published by ISO [1] establishes general rules for evaluating and expressing uncertainty for a wide range of measurements. The guide was interpreted for analytical chemistry by EURACHEM in 1995 [2]. The approach described in the GUM requires the identification of all possible sources of uncertainty associated with the procedure; the estimation of their magnitude from either experimental or published data; and the combination of these individual uncertainties to give standard and expanded uncertainties for the procedure as a whole. Some applications of this approach to analytical chemistry have been published [3, 4]. However, the GUM principles are significantly different from the methods currently used in analytical chemistry for estimating uncertainty [5–8] which generally make use of "whole method" performance parameters, such as precision and recovery, obtained during in-house method validation studies or during method development and collaborative study [9–11]. We have previously described a strategy for reconciling the information requirements of formal (i.e. GUM) measurement uncertainty principles with the data generated from method validation studies [12–14]. The approach involves a detailed analysis of the factors influencing the result using cause and effect analysis [15]. This results in a structured list of the possible sources of uncertainty associated with the method. The list is then simplified and reconciled with existing experimental and other data. We now report the application of this approach in the form of a protocol for the estimation of measurement uncertainty from validation studies [16]. This paper outlines the key stages in the protocol and discusses the use of data from trueness and ruggedness studies in detail. The practical application of the protocol will be described in Part 2 with reference to a high performance liquid chromatography (HPLC) procedure for the determination of markers in road fuel [17].

Principles of approach

The stages in the uncertainty estimation process are illustrated in Fig. 1. An outline of the procedure discussed in the protocol is presented in Fig. 2. The first stage of the procedure is the identification of sources of uncertainty for the method. Once the sources of uncertainty have been identified they require evaluation. The main tools for doing this are precision, trueness (or bias) and ruggedness studies. The aim is to account for as many sources of uncertainty as possible during the precision and trueness studies. Any remaining sources of uncertainty are then evaluated either from existing data (e.g. calibration certificates, published data, previous studies, etc.) or via ruggedness studies. Note that it may not be necessary to evaluate every source of uncertainty in detail, if the analyst has evidence to suggest that some are insignificant. Indeed, the EURACHEM Guide states that unless there are a large number of them, uncertainty components which are less than one-third of the largest need not be evaluated in detail. Finally, the individual uncertainty components for the method are combined to give standard and expanded uncertainties for the method as a whole. The use of data from trueness and ruggedness studies in uncertainty estimation is discussed in more detail below.

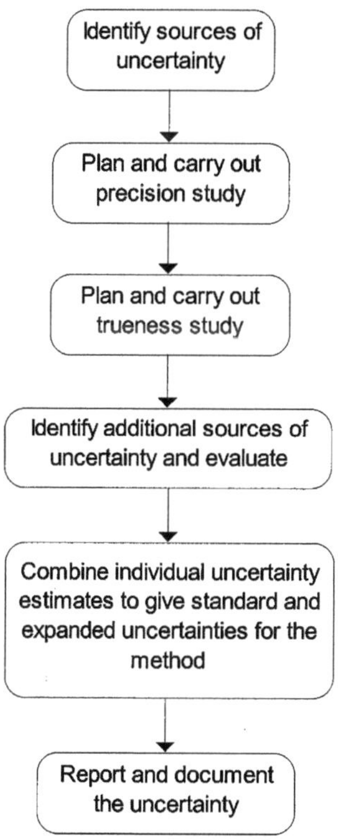

Fig. 1 Flow chart summarising the uncertainty estimation process

Trueness studies

In developing the protocol, the trueness of a method was considered in terms of recovery, i.e. the ratio of the observed value to the expected value. The evaluation of uncertainties associated with recovery is discussed in detail elsewhere [18, 19]. In general, the recovery, R, for a particular sample is considered as comprising three components:

- $\bar{R}_m$ is an estimate of the mean method recovery obtained from, for example, the analysis of a CRM or a spiked sample. The uncertainty in $\bar{R}_m$ is composed of the uncertainty in the reference value (e.g. the uncertainty in the certified value of a reference material) and the uncertainty in the observed value (e.g. the standard deviation of the mean of replicate analyses).
- R_s is a correction factor to take account of differences in the recovery for a particular sample compared to the recovery observed for the material used to estimate $\bar{R}_m$.
- R_{rep} is a correction factor to take account of the fact that a spiked sample may behave differently to a real sample with incurred analyte.

These three elements are combined multiplicatively to give an estimate of the recovery for a particular sample, R, and its uncertainty, $u(R)$:

$$R=\bar{R}_m \times R_s \times R_{\text{rep}}, \tag{1}$$

$$u(R)=R\times\sqrt{\left(\frac{u(\bar{R}_m)}{\bar{R}_m}\right)^2+\left(\frac{u(R_s)}{R_s}\right)^2+\left(\frac{u(R_{rep})}{R_{rep}}\right)^2}. \tag{2}$$

$\bar{R}_m$ and $u(\bar{R}_m)$ are calculated using Eq. (3) and Eq. (4):

$$\bar{R}_m=\frac{\bar{C}_{obs}}{C_{RM}}, \tag{3}$$

$$u(\bar{R}_m)=\bar{R}_m\times\sqrt{\left(\frac{s_{obs}}{\bar{C}_{obs}}\right)^2+\left(\frac{u(C_{RM})}{C_{RM}}\right)^2}, \tag{4}$$

where $\bar{C}_{obs}$ is the mean of the replicate analyses of the reference material (e.g. CRM or spiked sample), s_{obs} is the standard deviation of the mean of the results, C_{RM} is the concentration of the reference material and $u(C_{RM})$ is the standard uncertainty in the concentration of the reference material. To determine the contribu-

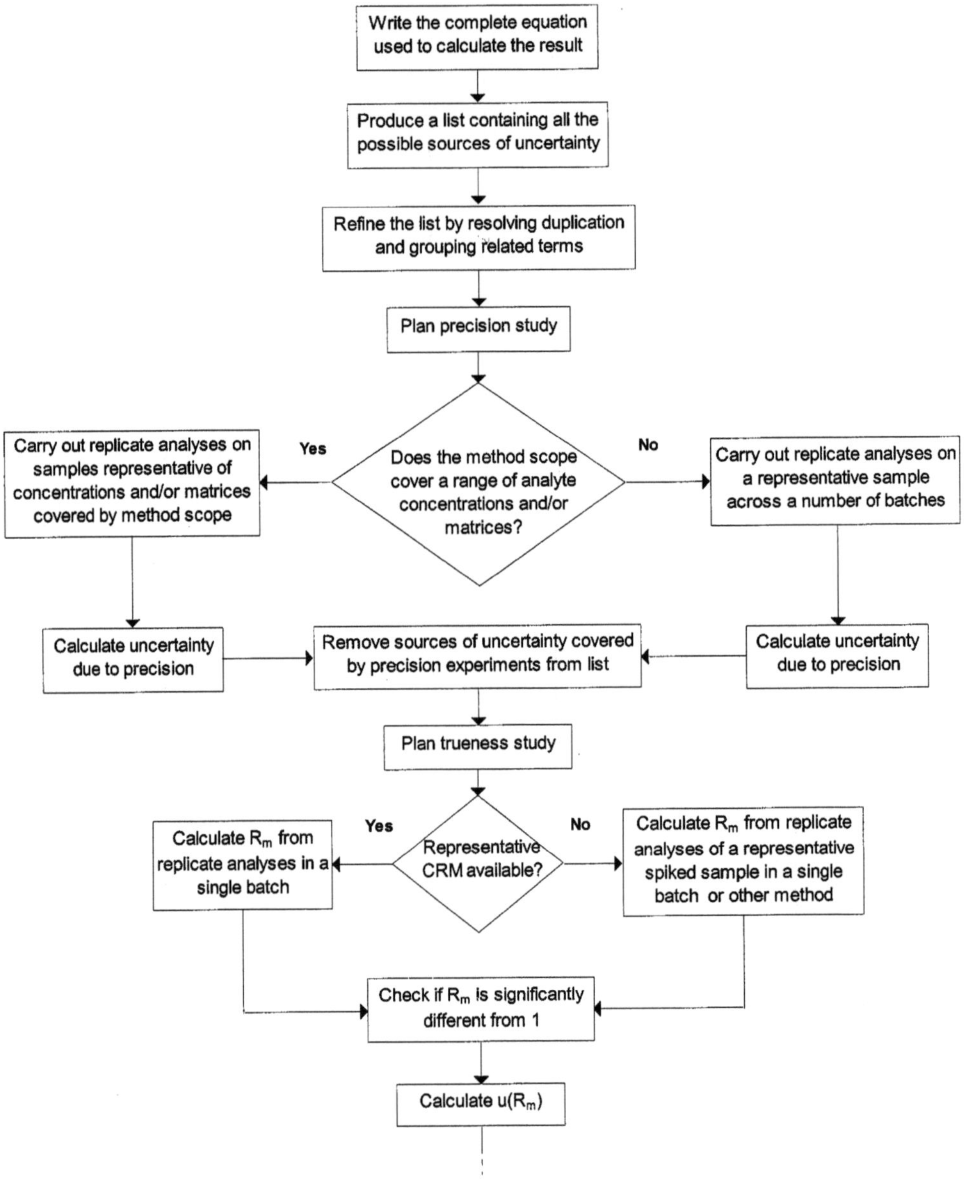

Fig. 2 Flow chart illustrating the stages in the method validation/measurement uncertainty protocol

tion of $\bar{R}_m$ to the combined uncertainty for the method as a whole, the estimate is compared with 1, using an equation of the form:

$$t = \frac{|1-\bar{R}_m|}{u(\bar{R}_m)} . \tag{5}$$

To determine whether $\bar{R}_m$ is significantly different from 1, the calculated value of t is compared with the coverage factor, $k=2$, which will be used to calculate the expanded uncertainty [19]. A t value greater than 2 suggests that $\bar{R}_m$ is significantly different from 1. However, if in the normal application of the method, no correction is made to take account of the fact that the method recovery is significantly different from 1, the uncertainty associated with $\bar{R}_m$ must be increased to take account of this uncorrected bias. The relevant equation is:

$$u(\bar{R}_m)' = \sqrt{\left(\frac{1-\bar{R}_m}{k}\right)^2 + u(\bar{R}_m)^2} . \tag{6}$$

A special case arises when an empirical method is being studied. In such cases, the method defines the measurand (e.g. dietary fibre, extractable cadmium from ceramics). The method is considered to define the true value and is, by definition, unbiased. The presumption is that $\bar{R}_m$ is equal to 1 and that the only uncertainty is that associated with the laboratory's particular application of the method. In some cases, a reference ma-

Fig. 2 Continued

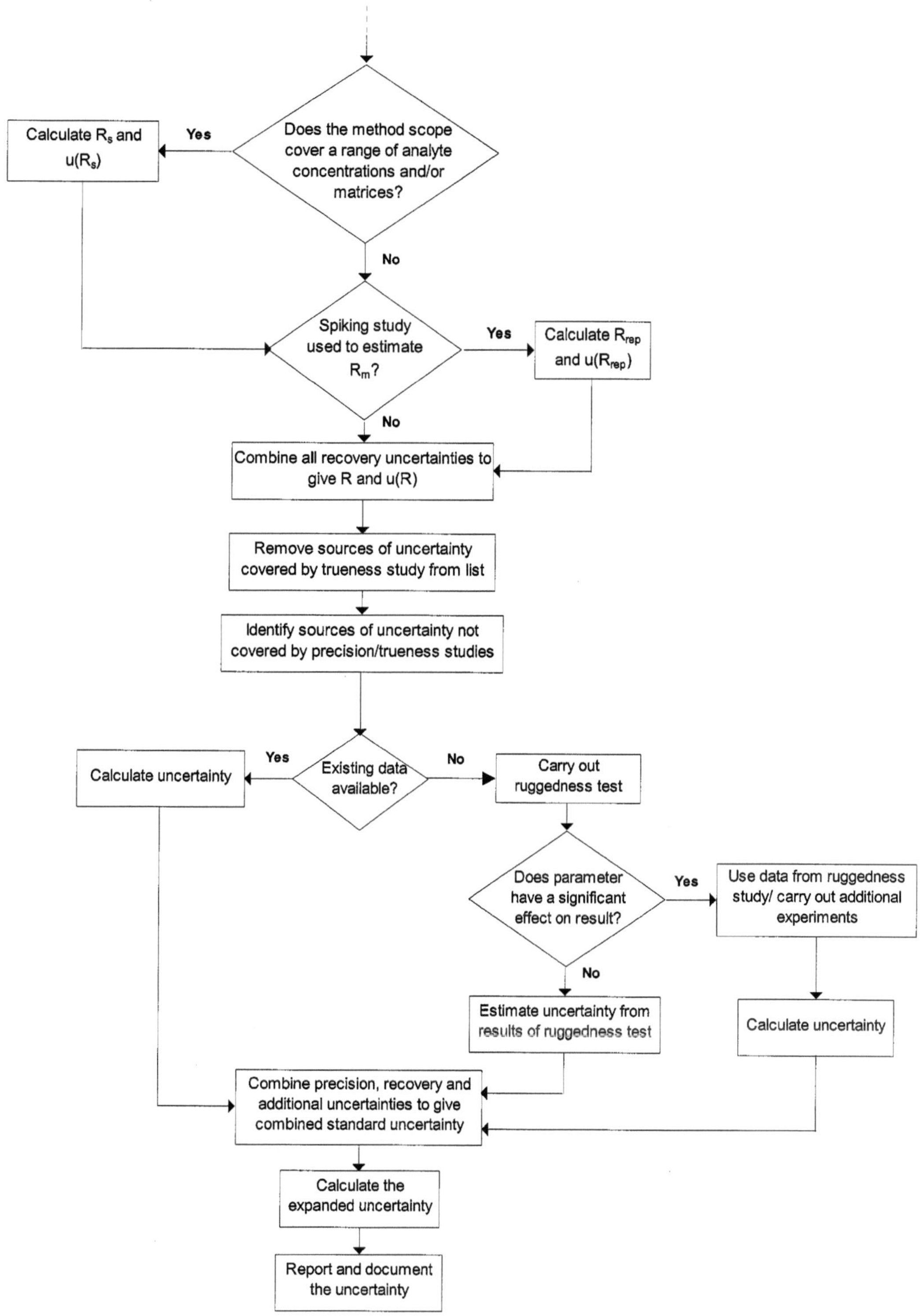

terial certified for use with the method may be available. Where this is so, a bias study can be carried out and the results treated as discussed above. If there is no relevant reference material, it is not possible to estimate the uncertainty associated with the laboratory bias. There will still be uncertainties associated with bias, but they will be associated with possible bias in the temperatures, masses, etc. used to define the method. In such cases it will normally be necessary to consider these individually.

Where the method scope covers a range of sample matrices and/or analyte concentrations, an additional uncertainty term R_s is required to take account of differences in the recovery of a particular sample type, compared to the material used to estimate $\bar{R}_m$. This can be evaluated by analysing a representative range of spiked samples, covering typical matrices and analyte concentrations, in replicate. The mean recovery for each sample type is calculated. R_s is normally assumed to be equal to 1. However, there will be an uncertainty associated with this assumption, which appears in the spread of mean recoveries observed for the different spiked samples. The uncertainty, $u(R_s)$, is therefore taken as the standard deviation of the mean recoveries for each sample type.

When a spiked sample, rather than a matrix reference material, has been used to estimate $\bar{R}_m$ it may be necessary to consider R_{rep} and its uncertainty. In general, R_{rep} is assumed to equal 1, indicating that the recovery observed for a spiked sample is truly representative of that for the incurred analyte. The uncertainty, $u(R_{rep})$, is a measure of the uncertainty associated with that assumption. In some cases it can be argued that a spike is a good representation of a real sample, for example in liquid samples where the analyte is simply dissolved in the matrix; $u(R_{rep})$ can therefore be assumed to be small. In other cases there may be reason to believe that a spiked sample is not a perfect model for a test sample and $u(R_{rep})$ may be a significant source of uncertainty. The evaluation of $u(R_{rep})$ is discussed in more detail elsewhere [18].

Evaluation of other sources of uncertainty

An uncertainty evaluation must consider the full range of variability likely to be encountered during application of the method. This includes parameters relating to the sample (analyte concentration, sample matrix) as well as experimental parameters associated with the method (e.g. temperature, extraction time, equipment settings, etc.). Sources of uncertainty not adequately covered by the precision and trueness studies require separate evaluation. There are three main sources of information: calibration certificates and manufacturers' specifications, data published in the literature and specially designed experimental studies. One efficient method of experimental study is ruggedness testing, discussed below.

Ruggedness studies

Ruggedness tests are a useful way of investigating simultaneously the effect of several experimental parameters on method performance. The experiments are based on the ruggedness testing procedure described in the *Statistical Manual of the AOAC* [20]. Such experiments result in an observed difference, D_{x_i}, for each parameter studied which represents the change in result due to varying that parameter. The parameters are tested for significance using a Student's t-test of the form [21]:

$$t=\frac{\sqrt{n}\times D_{x_i}}{\sqrt{2}\times s}, \tag{7}$$

where s is the estimate of the method precision, n is the number of experiments carried out at each level for each parameter ($n=4$ for a seven-parameter Plackett-Burman experimental design), and D_{x_i} is the difference calculated for parameter x_i. The values of t calculated using Eq. (7) are compared with the appropriate critical values of t at 95% confidence. Note that the degrees of freedom for t_{crit} relate to the degrees of freedom for the precision estimate used in the calculation of t. For parameters identified as having no significant effect on the method performance, the uncertainty in the final result y due to parameter x_i, $u(y(x_i))$, is calculated using Eq. (8):

$$u(y(x_i))=\frac{\sqrt{2}\times t_{\text{crit}}\times s}{\sqrt{n}\times 1.96}\times\frac{\delta_{\text{real}}}{\delta_{\text{test}}}, \tag{8}$$

where δ_{real} is the change in the parameter which would be expected when the method is operating under control in routine use and δ_{test} is the change in the parameter that was specified in the ruggedness study. In other words, the uncertainty estimate is based on the 95% confidence interval, converted to a standard deviation by dividing by 1.96 [1, 2]. The $\delta_{\text{real}}/\delta_{\text{test}}$ term is required to take account of the fact that the change in a parameter used in the ruggedness test may be greater than that observed during normal operation of the method. For parameters identified as having a significant effect on the method performance, a first estimate of the uncertainty can be calculated as follows:

$$u(y(x_i))=u(x_i)\times c_i, \tag{9}$$

$$c_i=\frac{\textit{Observed change in result}}{\textit{Change in parameter}}, \tag{10}$$

where $u(x_i)$ is the uncertainty in the parameter and c_i is the sensitivity coefficient.

Fig. 3 Contributions to the measurement uncertainty for the determination of CI solvent red 24, quinizarin and CI solvent yellow 124 in fuel oil

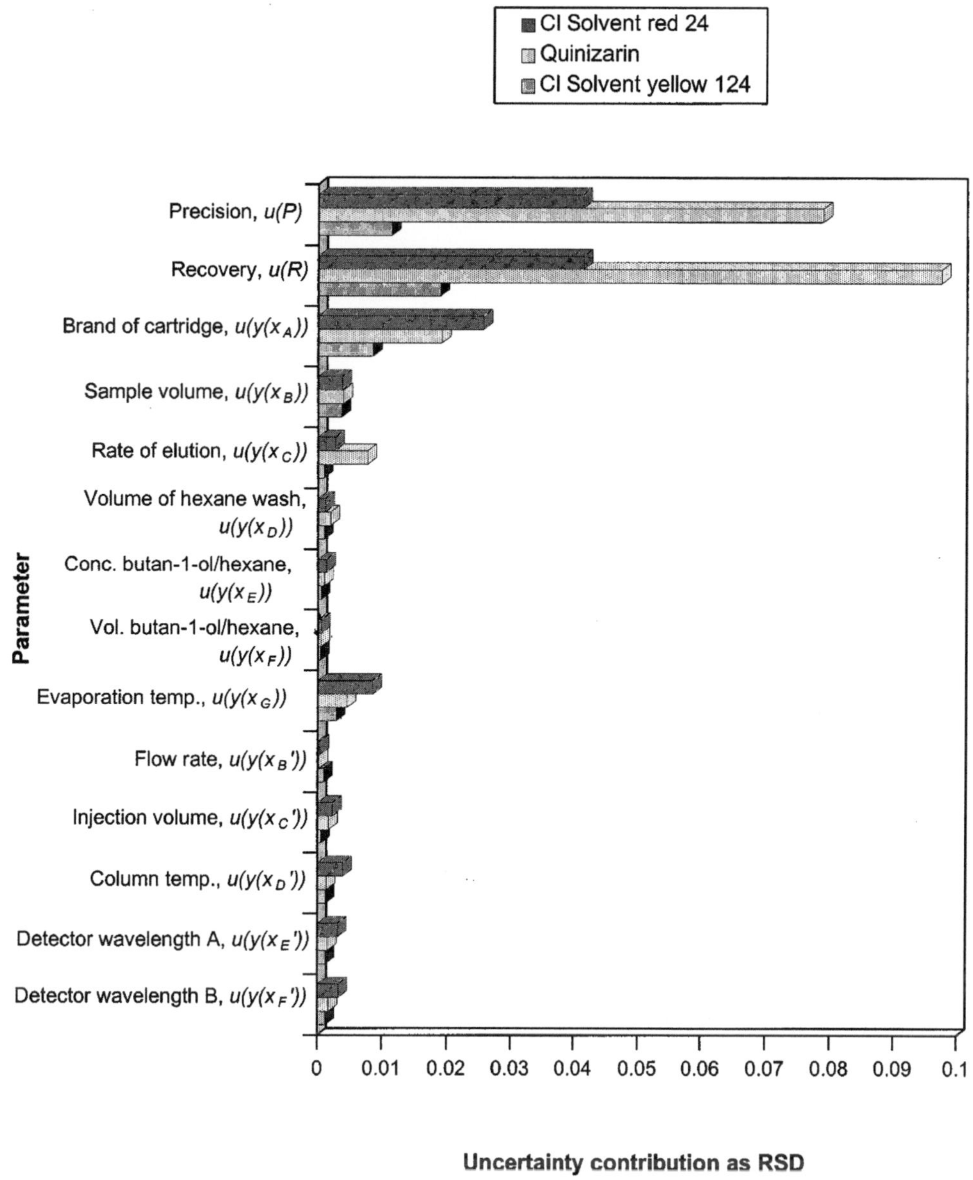

The estimates obtained by applying Eqs. 8–10 are intended to give a first estimate of the measurement uncertainty associated with a particular parameter. If such estimates of the uncertainty are found to be a significant contribution to the overall uncertainty for the method, further study of the effect of the parameters is advised, to establish the true relationship between changes in the parameter and the result of the method. However, if the uncertainties are found to be small compared to other uncertainty components (i.e. the uncertainties associated with precision and trueness) then no further study is required.

Calculation of combined measurement uncertainty for the method

The individual sources of uncertainty, evaluated through the precision, trueness, ruggedness and other studies are combined to give an estimate of the standard uncertainty for the method as a whole. Uncertainty contributions identified as being proportional to analyte concentration are combined using Eq. (11):

$$\frac{u(y)}{y}=\sqrt{\left(\frac{u(p)}{p}\right)^2+\left(\frac{u(q)}{q}\right)^2+\left(\frac{u(r)}{r}\right)^2+\dots}\,, \qquad (11)$$

where the result y is affected by parameters p, q, r.... which each have uncertainties $u(p)$, $u(q)$, $u(r)$.... Uncertainty contributions identified as being independent of analyte concentration are combined using Eq. (12):

$$u(y)' = \sqrt{u(p)^2 + u(q)^2 + u(r)^2 + \ldots} \tag{12}$$

The combined uncertainty in the result at a concentration y' is calculated as follows:

$$u(y') = \sqrt{(u(y)')^2 + \left(y' \times \frac{u(y)}{y}\right)^2} \tag{13}$$

Discussion and conclusions

We have developed a protocol which describes how data generated from experimental studies commonly undertaken for method validation purposes can be used in measurement uncertainty evaluation. The main experimental studies required are for the evaluation of precision and trueness. These should be planned so as to cover as many of the possible sources of uncertainty identified for the method as possible. Any remaining sources are considered separately. If there is evidence to suggest that they will be small compared to the uncertainties associated with precision and trueness, then no further study is required. However for uncertainty components where no prior information is available, further experimental study will be required. One useful approach is ruggedness testing which allows the evaluation of a number of sources of uncertainty simultaneously. It should be noted that ruggedness testing really only gives a first estimate of uncertainty contributions. Further study is recommended to refine the estimates for any sources of uncertainty which appear to be a significant contribution to the total uncertainty.

The main disadvantage of this approach is that it may not readily reveal the main sources of uncertainty for a particular method. In previous studies we have typically found the uncertainty budget to be dominated by the precision and trueness terms [14]. In such cases, if the combined uncertainty for the method is too large, indicating that the method requires improvement, further study may be required to identify the stages in the method which contribute most to the uncertainty. However, the approach detailed here will allow the analyst to obtain, relatively quickly, a sound estimate of measurement uncertainty, with minimum experimental work beyond that required for method validation.

We have applied this protocol to the evaluation of the measurement uncertainty for a method for the determination of three markers (CI solvent red 24, CI solvent yellow 124 and quinizarin (1,4-dihydroxyanthraquinone)) in road fuel. The method requires the extraction of the markers from the sample matrix by solid phase extraction, followed by quantification by HPLC with diode array detection. The uncertainty evaluation involved four experimental studies which were also required as part of the method validation. The studies were precision, trueness (evaluated via the analysis of spiked samples) and ruggedness tests of the extraction and HPLC stages. The experiments and uncertainty calculations are described in detail in Part 2. A summary of the uncertainty budget for the method is presented in Fig. 3.

Acknowledgements The work described in this paper was supported under contract with the Department of Trade and Industry as part of the United Kingdom National Measurement System Valid Analytical Measurement (VAM) Programme.

References

1. ISO (1993) Guide to the expression of uncertainty in measurement. ISO, Geneva
2. EURACHEM (1995) Quantifying uncertainty in analytical measurement. Laboratory of the Government Chemist (LGC), London
3. Pueyo M, Obiols J, Vilalta E (1996) Anal Commun 33:205–208
4. Williams A (1993) Anal Proc 30:248–250
5. Analytical Methods Committee (1995) Analyst 120:2303–2308
6. Ellison SLR (1997) In: Ciarlini P, Cox MG, Pavese F, Tichter D (eds) Advanced mathematical tools in metrology III. World Scientific, Singapore
7. Ellison SLR, Williams A (1998) Accred Qual Assur 3:6–10
8. Ríos A, Valcárcel M (1998) Accred Qual Assur 3:14–29
9. IUPAC (1988) Pure Appl Chem 60:885
10. AOAC (1989) J Assoc Off Anal Chem 72:694–704
11. ISO 5725:1994 (1994) Accuracy (trueness and precision) of measurement methods and results. ISO, Geneva
12. Ellison SLR, Barwick VJ (1998) Accred Qual Assur 3:101–105
13. Ellison SLR, Barwick VJ (1998) Analyst 123:1387–1392
14. Barwick VJ, Ellison SLR (1998) Anal Comm 35:377–383
15. ISO 9004-4:1993 (1993) Total quality management Part 2: Guidelines for quality improvement. ISO, Geneva
16. Barwick VJ, Ellison SLR (1999) Protocol for uncertainty evaluation from validation data. VAM Technical Report No. LGC/VAM/1998/088, available on LGC website at www.lgc.co.uk
17. Barwick VJ, Ellison SLR, Rafferty MJQ, Gill RS (1999) Accred Qual Assur
18. Barwick VJ, Ellison SLR (1999) Analyst 124:981–990
19. Ellison SLR, Williams A (1996) In: Parkany M (ed) The use of recovery factors in trace analysis. Royal Society of Chemistry, Cambridge
20. Youden WJ, Steiner EH (1975) Statistical manual of the association of official analytical chemists. Association of Official Analytical Chemists (AOAC), Arlington, Va.
21. Vander Heyden Y, Luypaert K, Hartmann C, Massart DL, Hoogmartens J, De Beer J (1995) Anal Chim Acta 312:245–262

Accred Qual Assur (2000) 5:104–113

Vicki J. Barwick
Stephen L.R. Ellison
Mark J.Q. Rafferty
Rattanjit S. Gill

The evaluation of measurement uncertainty from method validation studies

Part 2: The practical application of a laboratory protocol

Received: 10 April 1999
Accepted: 24 September 1999

V.J. Barwick (✉) · S.L.R. Ellison
M.J.Q. Rafferty · R.S. Gill
Laboratory of the Government Chemist,
Queens Road, Teddington, Middlesex,
TW11 0LY, UK
e-mail: vjb@lgc.co.uk,
Tel.: +44-20-8943 7421,
Fax: +44-20-8943 2767

Abstract A protocol has been developed illustrating the link between validation experiments and measurement uncertainty evaluation. The application of the protocol is illustrated with reference to a method for the determination of three markers (CI solvent red 24, quinizarin and CI solvent yellow 124) in fuel oil samples. The method requires the extraction of the markers from the sample matrix by solid phase extraction followed by quantification by high performance liquid chromatography (HPLC) with diode array detection. The uncertainties for the determination of the markers were evaluated using data from precision and trueness studies using representative sample matrices spiked at a range of concentrations, and from ruggedness studies of the extraction and HPLC stages.

Key words Measurement uncertainty · Method validation · Precision · Trueness · Ruggedness · High performance liquid chromatography

Introduction

In Part 1 [1] we described a protocol for the evaluation of measurement uncertainty from validation studies such as precision, trueness and ruggedness testing. In this paper we illustrate the application of the protocol to a method developed for the determination of the dyes CI solvent red 24 and CI solvent yellow 124, and the chemical marker quinizarin (1,4-dihydroxyanthraquinone) in road fuel. The analysis of road fuel samples suspected of containing rebated kerosene or rebated gas oil is required as the use of rebated fuels as road fuels or extenders to road fuels is illegal. To prevent illegal use of rebated fuels, HM Customs and Excise require them to be marked. This is achieved by adding solvent red 24, solvent yellow 124 and quinizarin to the fuel. A method for the quantitation of the markers was developed in this laboratory [2]. Over a period of time the method had been adapted to improve its performance and now required re-validation and an uncertainty estimate. This paper describes the experiments undertaken and shows how the data were used in the calculation of the measurement uncertainty.

Experimental

Outline of procedure for the determination of CI solvent red 24, CI solvent yellow 124 and quinizarin in fuel samples

Extraction procedure

The sample (10 ml) was transferred by automatic pipette to a solid phase extraction cartridge containing 500 mg silica. The cartridge was drained under vacuum until the silica bed appeared dry. The cartridge was then washed under vacuum with 10 ml hexane to remove residual oil. The markers were eluted from the cartridge under gravity with 10 ml butan-1-ol in hexane (10% v/v). The eluent was collected in a glass specimen vial and evaporated to dryness by heating to 50 °C under an air stream. The residue was dissolved in acetonitrile (2.5 ml) and the resulting solution placed in an ultrasonic bath for 5 min. The solution was then passed through a 0.45 μm filter prior to analysis by high performance liquid chromatography (HPLC).

HPLC conditions

The samples (50 μl) were analysed on a Hewlett Packard 1050 DAD system upgraded with a 1090 DAD optical bench. The column was a Luna 5 μm phenyl-hexyl, 250 mm × 4.6 mm maintained at 30 °C. The flow rate was 1 ml min^{-1} using a gradient elution of acetonitrile and water as follows:

Time (min)	0	3	4	5	9	10	20	21	23
Water	40	40	30	10	10	2	2	40	40
% Acetonitrile	60	60	70	90	90	98	98	60	60

Calibration was by means of a single standard in acetonitrile containing CI solvent red 24 and CI solvent yellow 124 at a concentration of approximately 20 mg l^{-1} and quinizarin at concentration of approximately 10 mg l^{-1}. CI solvent red 24 and quinizarin were quantified using data (peak areas) recorded on detector channel B (500 nm), whilst CI solvent yellow 124 was quantified using data recorded on detector channel A (475 nm). The concentration of the analyte, C in mg l^{-1}, was calculated using Eq. (1):

$$C = \frac{A_S \times V_F \times C_{STD}}{A_{STD} \times V_S}, \tag{1}$$

where A_S is the peak area recorded for the sample solution, A_{STD} is the peak area recorded for the standard solution, V_F is the final volume of the sample solution (ml), V_S is the volume of the sample taken for analysis (ml) and C_{STD} is the concentration of the standard solution (mg l^{-1}).

Experiments planned for validation and uncertainty estimation

A cause and effect diagram [3–5] illustrating the main parameters controlling the result of the analysis is presented in Fig. 1. Note that uncertainties associated with sampling are outside the scope of this study, as the uncertainty was required for the sample as received in the laboratory. The uncertainty contribution from sub-sampling the laboratory sample is represented by the "inhomogeneity" branch in Fig. 1. Initially, two sets of experiments were planned – a precision study and a trueness study. These were planned so as to cover as many sources of uncertainty as possible. Parameters not adequately covered by these experiments (i.e. not varied representatively) were evaluated separately using ruggedness tests or existing published data. Whilst these studies are required for the method validation process, it should be noted that they do not form a complete validation study [6].

Fig. 1 Cause and effect diagram illustrating sources of uncertainty for the method for the determination of markers in fuel oil

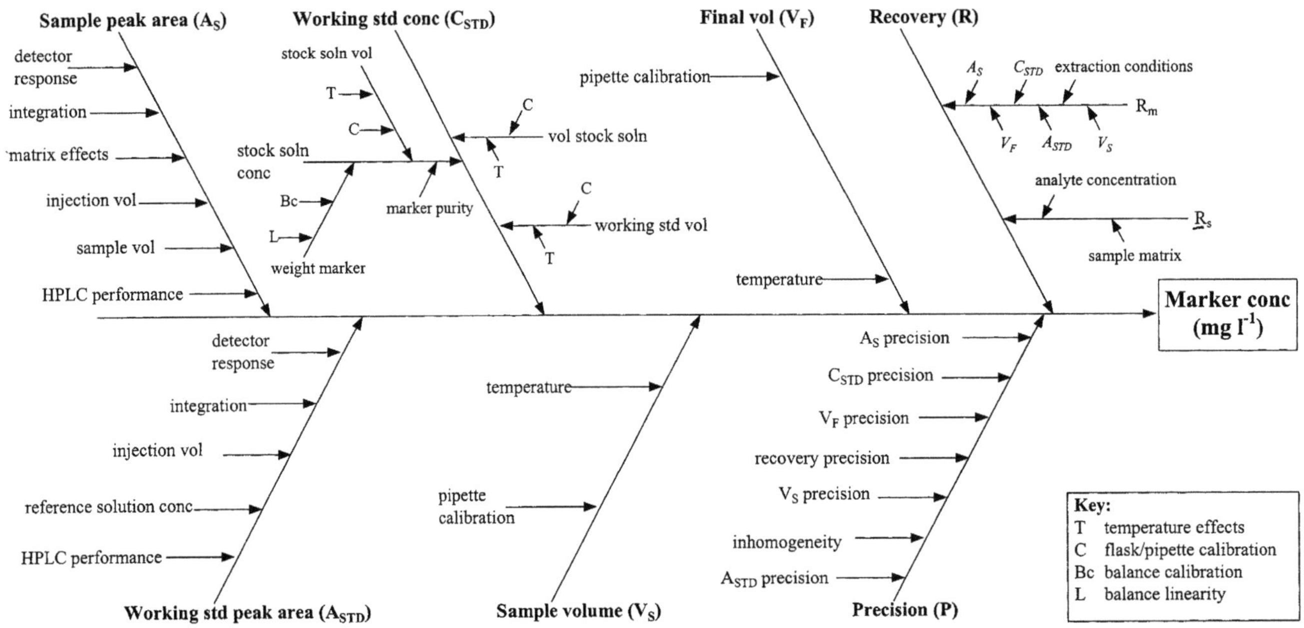

Precision experiments

Samples of 3 unmarked fuel oils (A–C) were fortified with CI solvent red 24 at concentrations of 0.041, 1.02, 2.03, 3.05 and 4.06 mg l^{-1}; quinizarin at concentrations of 0.040, 0.498, 0.996, 1.49 and 1.99 mg l^{-1}; and CI solvent yellow 124 at concentrations of 0.040, 1.20, 2.40, 3.99 and 4.99 mg l^{-1} to give a total of 15 fortified samples. Oil B was a diesel oil, representing a sample of typical viscosity. Oil A was a kerosene and oil C was a lubricating oil. These oils are respectively less viscous and more viscous than oil B.

Initially, 12 sub-samples of oil B with a concentration of 2.03 mg l^{-1} CI solvent red 24, 0.996 mg l^{-1} quinizarin and 2.40 mg l^{-1} CI solvent yellow 124 were analysed. The extraction stage was carried out in two batches of six on consecutive days. The markers in all 12 sub-samples were quantified in a single HPLC run, with the order of the analysis randomised. This study was followed by the analysis, in duplicate, of all 15 samples. The sample extracts were analysed in three separate HPLC runs such that the duplicates for each sample were in different runs. For each HPLC run a new standard and a fresh batch of mobile phase was prepared.

In addition, the results obtained from the replicate analysis of a sample of BP diesel, prepared for the trueness study (see below), were used in the estimate of uncertainty associated with method precision.

Trueness experiments

No suitable CRM was available for the evaluation of recovery. The study therefore employed representative samples of fuel oil spiked with the markers at the required concentrations. To obtain an estimate of $\bar{R}_m$ and its uncertainty, a 2-l sample of unmarked BP diesel was spiked with standards in toluene containing CI solvent red 24, quinizarin and CI solvent yellow 124 at concentrations of 0.996 mg ml^{-1}, 1.02 mg ml^{-1} and 1.97 mg ml^{-1}, respectively, to give concentrations in the diesel of 4.06 mg l^{-1}, 1.99 mg l^{-1} and 4.99 mg l^{-1}, respectively. A series of 48 aliquots of this sample were analysed in 3 batches of 16. The estimate of R_s and its uncertainty, $u(R_s)$, was calculated from these results plus the results from the analysis of the samples used in the precision study.

Evaluation of other sources of uncertainty: Ruggedness test

The effects of parameters associated with the extraction/clean-up stages and the HPLC quantification stage were studied in separate experiments. The parameters studied for the extraction/clean-up stage of the method and the levels chosen are shown in Table 1a. The ruggedness test was applied to the matrix B (diesel oil) sample containing 2.03 mg l^{-1} CI solvent red 24, 0.996 mg l^{-1} quinizarin and 2.40 mg l^{-1} CI solvent yellow 124 used in the precision study. The eight experiments were carried out over a short period of time and the resulting sample extracts were analysed in a single HPLC run. The HPLC parameters investigated and the levels chosen are given in Table 1b. For this set of experiments a single extract of the matrix B (diesel oil) sample, obtained under normal method conditions, was used. The extract and a standard were run under each set of conditions required by the ruggedness test. The effect of variations in the parameters was monitored by calculating the concentration of the markers observed under each set of parameters, using the appropriate standard.

Results and uncertainty calculations

Precision study

The results from the precision studies are summarised in Table 2. Estimates for the standard deviation for a single result were obtained from the results of the duplicate analyses of the 15 samples, by taking the standard deviation of the differences between the pairs and dividing by $\sqrt{2}$. Estimates of the relative standard deviations were obtained by treating the normalised differences in the same way [7]. The results from the analysis of the BP diesel sample represented three batches of 16 replicate analyses. An estimate of the total precision (i.e. within and between batch variation) was obtained via ANOVA [8]. The precision estimates cover different sources of variability in the method. The estimates obtained from the duplicate samples and the BP oil sample cover batch to batch variability in the extraction and HPLC stages of the method (including the preparation of new standards and mobile phase). The estimate obtained from matrix B does not cover batch to batch variability in the HPLC procedure as all the replicates were analysed in a single HPLC run. The precision studies also cover the uncertainty associated with sample inhomogeneity as they involved the analysis of a number of sub-samples taken from the bulk.

CI solvent red 24

No significant difference was observed (F-tests, 95% confidence) between the three estimates obtained for the relative standard deviation (0.0323, 0.0289 and 0.0414). However, the test was borderline and across the range studied (0.04 mg l^{-1} to 4 mg l^{-1}) the method

Table 1 Results from the ruggedness testing of the procedure for the determination of CI solvent red 24, quinizarin and CI solvent yellow 124 in fuel oil. **a** Ruggedness testing of the extraction/clean-up procedure

Parameter	Values				$\delta_{real}/\delta_{test}$	$u(x_i)$	CI solvent red 24			Quinizarin			CI solvent yellow 124		
							D_{x_i} (mg l^{-1})	c_i	$u(y(x_i))$ (mg l^{-1})	D_{x_i} (mg l^{-1})	c_i	$u(y(x_i))$ (mg l^{-1})	D_{x_i} (mg l^{-1})	c_i	$u(y(x_i))$ (mg l^{-1})
Brand of silica cartridges	A	Varian	a	Waters	1/1[a]		0.00750*	–	0.0493	−0.00750*	–	0.0174	−0.00250*	–	0.0199
Sample volume	B	10 ml	b	12 ml	–	0.04 ml	−0.353	0.176	0.00705	−0.180	0.090	0.00360	−0.423	0.212	0.00845
Rate of elution of oil with hexane	C	vacuum	c	gravity	1/10[a]	–	0.0275*	–	0.00493	0.070	–	0.0070[a]	−0.020*	–	0.00199
Volume of hexane wash	D	12 ml	d	8 ml	–	0.04 ml	0.213	0.0531	0.00213	0.176	0.0444	0.00177	0.225	0.0563	0.00225
Concentration of butan-1-ol/hexane	E	12%	e	8%	0.2% (v/v)/ 4% (v/v)		−0.0425*	–	0.00247	0.0175*	–	0.000868	−0.010*	–	0.0010
Volume 10% of butan-1-ol/hexane	F	12 ml	f	8 ml	0.08 ml/ 4 ml	0.04 ml	0.0625*	–	0.000986	−0.0050*	–	0.000347	0.080	0.020	0.00080
Evaporation temperature	G	50 °C	g	80 °C	10 °C/30 °C	2.89 °C	−0.0275*	–	0.0164	−0.0425	0.00142	0.00409	0.00750*	–	0.00663

[a] See text for explanation
*No significant effect at the 95% confidence level

Table 1b Ruggedness testing of the HPLC procedure

Parameter	Values				$\delta_{real}/\delta_{test}$	$u(x_i)$	CI solvent red 24			Quinizarin			CI solvent yellow 124		
							D_{x_i} (mg l^{-1})	c_i	$u(y(x_i))$ (mg l^{-1})	D_{x_i} (mg l^{-1})	c_i	$u(y(x_i))$ (mg l^{-1})	D_{x_i} (mg l^{-1})	c_i	$u(y(x_i))$ (mg l^{-1})
Type of acetonitrile in mobile phase	A′	Far-UV grade	a′	HPLC grade	–	a	−0.0748	a	a	−0.101	a	a	−0.0228*	a	a
Flow rate	B′	0.8 ml min^{-1}	b′	1.2 ml min^{-1}	–	0.00173 ml min^{-1}	−0.124	0.309	0.000535	0.0283	0.0707	0.000122	−0.0465	0.116	0.00020
Injection volume	C′	40 µl	c′	60 µl	–	0.75 µl	0.115	0.00576	0.00432	−0.0406	0.00203	0.00152	0.0284	0.00142	0.00107
Column temperature	D′	25 °C	d′	35 °C	2 °C/10 °C	1 °C	−0.130	0.0130	0.00752	0.0201	0.00201	0.00116	−0.0233*	–	0.00282
Detector wavelength (A)	E′	465 nm	e′	485 nm	4 nm/20 nm	1.15 nm	0.104	0.00520	0.00598	0.0239	0.00120	0.00138	−0.0161*	–	0.00282
Degassing of mobile phase	F′	Degassed	f′	Not degassed	–	a	0.108	a	a	0.0641	a	a	0.00907*	a	a
Detector wavelength (B)	G′	490 nm	g′	510 nm	4 nm/ 20 nm	1.15 nm	0.105	0.00525	0.00604	−0.0112*	–	0.00154	0.0198*	–	0.00282

[a] See text for explanation
*No significant effect at the 95% confidence level

Table 2 Summary of data used in the estimation of $u(P)$

Analyte/Matrix	n	Mean ($mg\,l^{-1}$)	Standard deviation ($mg\,l^{-1}$)	Relative standard deviation
CI solvent red 24				
Matrix B	12	1.92	0.0621	0.0323
BP diesel	48[a]	3.88	0.112	0.0289
Matrices A–C	15[b]	–	0.0376	0.0414
Quinizarin				
Matrix B	11	0.913	0.0216	0.0236
BP diesel	48[a]	1.89	0.0256	0.0136
Matrices A–C	15[b]	–	0.0470	0.0788
CI solvent yellow 124				
Matrix B	12	2.35	0.0251	0.0107
BP diesel	48[a]	4.99	0.0618	0.0124
Matrices A–C	15[b]	-	0.0247	0.0464

[a] Standard deviation and relative standard deviation estimated from ANOVA of 3 sets of 16 replicates (see text)
[b] Standard deviation and relative standard deviation estimated from duplicate results (15 sets) for a range of concentrations and matrices (see text)

precision was approximately proportional to analyte concentration. It was decided to use the estimate of 0.0414 as the uncertainty associated with precision, $u(P)$, to avoid underestimating the precision for any given sample. This estimate was obtained from the analysis of different matrices and concentrations and is therefore likely to be more representative of the precision across the method scope.

Quinizarin

The estimates of the standard deviation and relative standard deviation were not comparable. In particular, the estimates obtained from the duplicate results were significantly different from the other estimates (F-tests, 95% confidence). There were no obvious patterns in the data so no particular matrix and/or concentration could be identified as being the cause of the variability. There was therefore no justification for removing any data and restricting the coverage of the uncertainty estimate, as in the case of CI solvent yellow 124 (see below). The results of the precision studies indicate that the method is more variable across different matrices and analyte concentrations for quinizarin than for the other markers. The uncertainty associated with the precision was taken as the estimate of the relative standard deviation obtained from the duplicate results, 0.0788. This estimate should ensure that the uncertainty is not underestimated for any given matrix or concentration (although it may result in an overestimate in some cases).

CI solvent yellow 124

There was no significant difference between the estimates of the relative standard deviation obtained for samples at concentrations of $2.4\,mg\,l^{-1}$ and $4.99\,mg\,l^{-1}$. However, the estimate obtained from the duplicate analyses was significantly greater than the other estimates. Inspection of that data revealed that the normalised differences observed for the samples at a concentration of $0.04\,mg\,l^{-1}$ were substantially larger than those observed at the other concentrations. Removing these data points gave a revised estimate of the relative standard deviation of 0.00903. This was in agreement with the other estimates obtained (F-tests, 95% confidence). The three estimates were therefore pooled to give a single estimate of the relative standard deviation of 0.0114. At present, the uncertainty estimate cannot be applied to samples with concentrations below $1.2\,mg\,l^{-1}$. Further study would be required to investigate in more detail the precision at these low levels.

Trueness study

Evaluation of $\bar{R}_m$ and $u(\bar{R}_m)$

The results are summarised in Table 3. In each case $\bar{R}_m$ was calculated using Eq. (2):

$$\bar{R}_m = \frac{\bar{C}_{obs}}{C_{RM}}, \tag{2}$$

where $\bar{C}_{obs}$ is the mean of the replicate analyses of the spiked sample and C_{RM} is the concentration of the

Table 3 Results from the replicate analysis of a diesel oil spiked with CI solvent red 24, quinizarin and CI solvent yellow 124

Analyte	Target concentration, C_{spike} ($mg\,l^{-1}$)	Mean, $\bar{C}_{obs}$ ($mg\,l^{-1}$)	Standard deviation of the mean, s_{obs} ($mg\,l^{-1}$)[a]
CI solvent red 24	4.06	3.88	0.0360
Quinizarin	1.99	1.89	0.00370
CI solvent yellow 124	4.99	4.99	0.0167

[a] Estimated from ANOVA of 3 groups of 16 replicates according to ISO 5725:1994 [9]

spiked sample. The uncertainty, $u(\bar{R}_m)$, was calculated using Eq. (3):

$$u(\bar{R}_m)=\bar{R}_m\times\sqrt{\left(\frac{s_{obs}}{\bar{C}_{obs}}\right)^2+\left(\frac{u(C_{RM})}{C_{RM}}\right)^2}, \quad (3)$$

where $u(C_{RM})$ is the standard uncertainty in its concentration of the spiked sample. The standard deviation of the mean of the results, s_{obs}, was estimated from ANOVA of the data according to Part 4 of ISO 5725:1994 [9].

Using information on the purity of the material used to prepare the spiked sample, and the accuracy and precision of the volumetric glassware and analytical balance used, the uncertainty in the concentration of CI solvent red 24 in the sample, $u(C_{RM})$, was estimated as 0.05 mg l^{-1}.[1] The uncertainties associated with the concentration of quinizarin and CI solvent yellow 124 were estimated as 0.025 mg l^{-1} and 0.062 mg l^{-1}, respectively. The relevant values are:

CI solvent red 24:	$\bar{R}_m=0.957$	$u(\bar{R}_m)=0.0148$
Quinizarin:	$\bar{R}_m=0.949$	$u(\bar{R}_m)=0.0121$
CI solvent yellow 124:	$\bar{R}_m=1.00$	$u(\bar{R}_m)=0.0129$

Applying Eq. (4):

$$t=\frac{|1-\bar{R}_m|}{u(\bar{R}_m)} \quad (4)$$

indicated that the estimates of $\bar{R}_m$ obtained for CI solvent red 24 and quinizarin were significantly different from 1.0 ($t>2$) [7, 10]. During routine use of the method, the results reported for test samples will not be corrected for incomplete recovery of the analyte. Equation (5) was therefore used to calculate an increased uncertainty for $\bar{R}_m$ to take account of the uncorrected bias:

$$u(\bar{R}_m)'=\sqrt{\left(\frac{1-\bar{R}_m}{k}\right)^2+u(\bar{R}_m)^2}, \quad (5)$$

$u(\bar{R}_m)'$ was calculated as 0.0262 for CI solvent red 24 and 0.0283 for quinizarin. The significance test for CI solvent yellow 124 indicated that $\bar{R}_m$ was not significantly different from 1.0. The uncertainty associated with $\bar{R}_m$ is the value of $u(\bar{R}_m)$ calculated above (i.e. 0.0129).

$u(R_s)$ is the standard deviation of the mean recoveries obtained for the samples analysed in the precision studies and the BP diesel sample used in the study of $\bar{R}_m$. This gave estimates of $u(R_s)$ of 0.0322 for CI solvent red 24, 0.0932 for quinizarin and 0.0138 for CI solvent yellow 124. The estimate for CI solvent yellow 124 only includes concentrations above 1.2 mg l^{-1}, for the reason discussed in the section on precision.

[1] Detailed information on the estimation of uncertainties of this type is given in Ref. [7].

Calculation of R and u(R)

The recovery, R, for a particular test sample and the corresponding uncertainty, $u(R)$, is calculated using Eqs. (6) and (7):

$$R=\bar{R}_m\times R_s\times R_{rep}, \quad (6)$$

$$u(R)=R\times\sqrt{\left(\frac{u(\bar{R}_m)}{\bar{R}_m}\right)^2+\left(\frac{u(R_s)}{R_s}\right)^2+\left(\frac{u(R_{rep})}{R_{rep}}\right)^2}. \quad (7)$$

In this study a spiked sample can be considered a reasonable representation of test samples of marked fuel oils. There is therefore no need to correct the estimates of $\bar{R}_m$ and $u(\bar{R}_m)$ by including the R_{rep} and $u(R_{rep})$ terms. Both $\bar{R}_m$ and R_s are assumed to be equal to 1. R is therefore also equal to 1. Combining the estimates of $u(\bar{R}_m)$ and $u(R_s)$, the uncertainty $u(R)$ was calculated as 0.0415 for CI solvent red 24, 0.0974 for quinizarin and 0.0187 for CI solvent yellow 124.

Ruggedness test of extraction/clean-up procedure

The results from the ruggedness study of the extraction/clean-up procedure are presented in Table 1a. The precision of the method for the analysis of the sample used in the ruggedness study had been estimated previously as 0.0621 mg l^{-1} ($\nu=11$) for CI solvent red 24, 0.0216 mg l^{-1} ($\nu=10$) for quinizarin and 0.0251 mg l^{-1} ($\nu=11$) for CI solvent yellow 124. Parameters were tested for significance using Eq. (8):

$$t=\frac{\sqrt{n}\times D_{x_i}}{\sqrt{2}\times s}, \quad (8)$$

where s is the estimate of the method precision, n is the number of experiments carried out at each level for each parameter ($n=4$ for a seven-parameter Plackett-Burman experimental design), and D_{x_i} is the difference calculated for parameter x_i [1, 11]. The degrees of freedom for t_{crit} relate to the degrees of freedom for the precision estimate used in the calculation of t.

The parameters identified as having no significant effect on method performance, at the 95% confidence level are highlighted in Table 1a. For these parameters the uncertainty in the final result was calculated using Eq. (9):

$$u(y(x_i))=\frac{\sqrt{2}\times t_{crit}\times s}{\sqrt{n}\times 1.96}\times\frac{\delta_{real}}{\delta_{test}}, \quad (9)$$

where δ_{real} is the change in the parameter which would be expected when the method is operating under control in routine use and δ_{test} is the change in the parameter that was specified in the ruggedness study. The estimates of δ_{real} are given in Table 1a. For parameter A, brand of silica cartridge, the conditions of the test (i.e. changing between two brands of cartridge) were considered representative of normal operation of the method. δ_{real} is therefore equal to δ_{test}. The effect of the rate of elution of oil by hexane from the cartridge was investigated by comparing the elution under a vacuum and with elution under gravity. In routine analyses, the oil will be eluted under vacuum. Variations in the vacuum applied from one extraction to another will affect the rate of elution of the oil and the amount of oil eluted. However, the effect of variations in the vacuum will be small compared to the effect of having no vacuum present. It can therefore be assumed that variations in the observed concentration of the markers, due to variability in the vacuum, will be small compared to the differences observed in the ruggedness test. As a first estimate, the effect of variation in the vacuum during routine application of the method was estimated as one-tenth of that observed during the ruggedness study. This indicated that the parameter was not a significant contribution to the overall uncertainty for CI solvent red 24 and CI solvent yellow 124, so no further study was required. The estimates of δ_{real} for the concentration and volume of butan-1-ol in hexane used to elute the column were based on the manufacturers' specifications and typical precision data for the volumetric flasks and pipettes used to prepare and deliver the solution.

For the parameters identified as having a significant effect on method performance, the uncertainty was calculated using Eqs. (10) and (11):

$$u(y(x_i)) = u(x_i) \times c_i, \tag{10}$$

$$c_i = \frac{\textit{Observed change in result}}{\textit{Change in parameter}}. \tag{11}$$

The estimates of the uncertainty in each parameter, $u(x_i)$, are given in Table 1a. The uncertainties associated with the sample volume, volume of hexane wash and volume of the 10% butan-1-ol/hexane solution were again based on the manufacturers' specifications and typical precision data for the volumetric flasks and pipettes used to prepare and deliver the solutions. The uncertainty in the evaporation temperature was based on the assumption that the temperature could be controlled to ±5 °C. This was taken as a rectangular distribution and converted to a standard uncertainty by dividing by $\sqrt{3}$ [7]. As discussed previously, the effect on the final result of variations in the vacuum when eluting the oil from the cartridge with hexane was estimated as one-tenth that observed in the ruggedness test.

The effects of all the parameters were considered to be proportional to the analyte concentration. The uncertainties were therefore converted to relative standard deviations by dividing by the mean of the results obtained from previous analyses of the sample under normal method conditions (see results for Matrix B in Table 2).

Ruggedness test of the HPLC procedure

The results from the ruggedness study of the HPLC procedure, and the values of δ_{real} and $u(x_i)$ used in the uncertainty calculations, are presented in Table 1b. Replicate analyses of a standard solution of the three markers gave the following estimates of the precision of the HPLC system at the concentration of the sample used in the study: CI solvent red 24, $s = 0.0363$ mg l^{-1} ($n = 69$); quinizarin, $s = 0.0107$ mg l^{-1} ($n = 69$); CI solvent yellow 124, $s = 0.0196$ mg l^{-1} ($n = 69$). Parameters were tested for significance, at 95% confidence, using Eq. (8). The uncertainties for parameters identified as having no significant effect on the method performance were calculated using Eq. (9). Based on information from manufacturers' specifications for HPLC systems, the uncertainty associated with the column temperature was estimated as ±1 °C, giving an estimate of δ_{real} of 2 °C. Again, based on manufacturers' specifications for DAD detectors, the uncertainty associated with the detector wavelengths was estimated as ±2 nm, giving a δ_{real} value of 4 nm.

The uncertainties due to significant parameters were estimated using Eqs. (10) and (11). Information in the literature suggests that a typical variation in flow rate is ±0.3% [12]. The uncertainty in the flow rate was therefore estimated as 0.00173 ml min^{-1}, assuming a rectangular distribution. Data in the literature gave 1.5% as a typical coefficient of variation for the volume delivered by an autosampler [13]. The uncertainty associated with the injection volume of 50 μl was therefore estimated as 0.75 μl.

Two remaining parameters merit further discussion; the type of acetonitrile used in the mobile phase and whether or not the mobile phase was degassed. The method was developed using HPLC grade acetonitrile. The ruggedness test indicated that changing to far-UV grade results in a lower recovery for all three analytes. The method protocol should therefore specify that for routine use, HPLC grade acetonitrile must be used. The ruggedness test also indicated that not degassing the mobile phase causes a reduction in recovery. The method was developed using degassed mobile phase, and the method protocol will specify that this must be the case during future use of the method. As these two parameters are being controlled in the method protocol, uncertainty terms have not been included.

The effects of all the parameters were considered to be proportional to the analyte concentration. The uncertainties were therefore converted to relative standard deviations by dividing by the mean of results obtained from previous analyses of the sample under normal method conditions (see results for Matrix B in Table 2).

Other sources of uncertainty

The precision and trueness studies were designed to cover as many of the sources of uncertainty as possible (see Fig. 1), for example, by analysing different sample matrices and concentration levels, and by preparing new standards and HPLC mobile phase for each batch of analyses. Parameters which were not adequately varied during these experiments, such as the extraction and HPLC conditions, were investigated in the ruggedness tests. There are however, a small number of parameters which were not covered by the above experiments. These generally related to the calibration of pipettes and balances used in the preparation of the standards and samples. For example, during this study the same pipettes were used in the preparation of all the working standards. Although the precision associated with the operation of the pipette is included in the overall precision estimate, the effect of the accuracy of the pipettes has not been included in the uncertainty budget so far. A pipette used to prepare the standard may typically deliver 0.03 ml above its nominal value. In the future a different pipette, or the same pipette

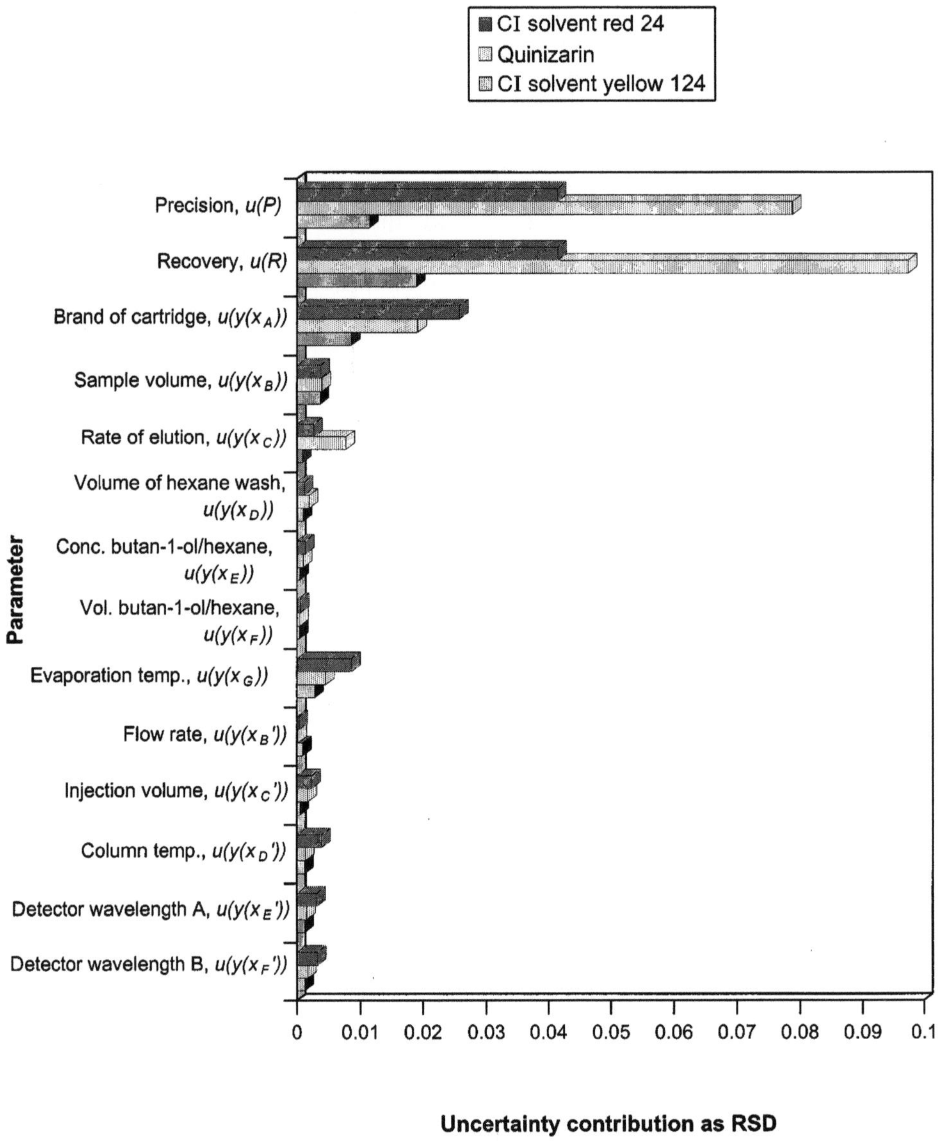

Fig. 2 Contributions as relative standard deviations (RSDs) to the measurement uncertainty for the determination of CI solvent red 24, quinizarin and CI solvent yellow 124 in fuel oil

after re-calibration, may deliver 0.02 ml below the nominal value. Since this possible variation is not already included in the uncertainty budget it should be considered separately. However, previous experience [14] has shown us that uncertainties associated with the calibration of volumetric glassware and analytical balances are generally small compared to other sources of uncertainty such as overall precision and recovery. Additional uncertainty estimates for these parameters have not therefore been included in the uncertainty budgets.

Calculation of measurement uncertainty

The contributions to the uncertainty budget for each of the analytes are illustrated in Fig. 2. In all cases the sources of uncertainty were considered to be proportional to analyte concentration. Using Eq. (12):

$$\frac{u(y)}{y}=\sqrt{\left(\frac{u(p)}{p}\right)^2+\left(\frac{u(q)}{q}\right)^2+\left(\frac{u(r)}{r}\right)^2+\ldots} \qquad (12)$$

the uncertainty in the final result, $u(y)$, was calculated as 0.065 for CI solvent red 24, 0.13 for quinizarin and 0.024 for CI solvent yellow 124, all expressed as relative standard deviations. The expanded uncertainties, calculated using a coverage factor of $k=2$ which gives a confidence level of approximately 95%, are 0.13, 0.26 and 0.048 for CI solvent red 24, quinizarin and CI solvent yellow 124, respectively.

Discussion

In the case of CI solvent red 24 and CI solvent yellow 124, the significant contributions to the uncertainty budget arose from overall precision and recovery, and the brand of the solid phase extraction cartridge used. If a reduction in the overall uncertainty of the method was required, useful approaches would be to specify a particular brand of cartridge in the method protocol, or to adopt matrix specific recovery corrections for test samples.

The combined uncertainty for quinizarin, which is significantly greater than that calculated for the other markers, is dominated by the precision and recovery terms. The results of the precision study indicated variable method performance across different matrices and analyte concentrations. The uncertainty, $u(R_s)$, associated with the variation in recovery from sample to sample was the major contribution to the recovery uncertainty, $u(R)$. This was due to the fact that the recoveries obtained for matrix B were generally higher than those obtained for matrices A and C. However, in this study, a single uncertainty estimate for all the matrices and analyte concentrations studied was required. It was therefore necessary to use "worst case" estimates of the uncertainties for precision and recovery to adequately cover all sample types. If this estimate was found to be unsatisfactory for future applications of the method, separate budgets could be calculated for individual matrices and concentration ranges.

Conclusions

We have developed a protocol which describes how data generated from experimental studies commonly undertaken for method validation purposes can be used in measurement uncertainty evaluation. This paper has illustrated the application of the protocol. In the example described, the uncertainty estimate for three analytes in different oil matrices was evaluated from three experimental studies, namely precision, recovery and ruggedness. These studies were required as part of the method validation, but planning the studies with uncertainty evaluation in mind allowed an uncertainty estimate to be calculated with little extra effort. A number of areas were identified where additional experimental work may be required to refine the estimates. However the necessary data could be generated by carrying out additional analyses alongside routine test samples. Again this would minimise the amount of laboratory effort required.

For methods which are already in routine use there may be historical validation data available which could be used, in the same way as illustrated here, to generate an uncertainty estimate. If no such data are available, the case study gives an indication on the type of experimental studies required. Again, with careful planning, it is often possible to undertake the studies alongside routine test samples.

Acknowledgments The work described in this paper was supported under contract with the Department of Trade and Industry as part of the United Kingdom National Measurement System Valid Analytical Measurement (VAM) Programme.

References

1. Barwick VJ, Ellison SLR (1999) Accred Qual Assur (in press)
2. May EM, Hunt DC, Holcombe DG (1986) Analyst 111:993–995
3. Ellison SLR, Barwick VJ (1998) Accred Qual Assur 3:101–105
4. Ellison SLR, Barwick VJ (1998) Analyst 123:1387–1392
5. ISO 9004-4:1993 (1993) Total quality management Part 2. Guidelines for quality improvement. ISO, Geneva, Switzerland
6. EURACHEM (1998) The fitness for purpose of analytical methods, a laboratory guide to method validation and related topics. Laboratory of the Government Chemist, London
7. EURACHEM (1995) Quantifying uncertainty in analytical measurement. Laboratory of the Government Chemist, London
8. Farrant TJ (1997) Practical statistics for the analytical scientist: a bench guide. Royal Society of Chemistry, Cambridge
9. ISO 5725:1994 (1994) Accuracy (trueness and precision) of measurement methods and results. ISO, Geneva, Switzerland
10. Ellison SLR, Williams A (1996) In: Parkanay M (ed) The use of recovery factors on trace analysis. Royal Society of Chemistry, Cambridge
11. Youden WJ, Steiner EH (1975) Statistical manual of the association of official analytical chemists. Association of Official Analytical Chemists, Arlington, Va
12. Brown PR, Hartwick RA (eds) (1989) High performance liquid chromatography. Wiley, New York
13. Dolan JW (1997) LC-GC International 10:418–422
14. Barwick VJ, Ellison SLR (1998) Anal Comm 35:377–383

Accred Qual Assur (2001) 6:31–36

Solveig Linko

Automated ion-selective measurement of lithium in serum. A practical approach to result-level verification in a two-way method validation

Received: 19 April 2000
Accepted: 26 July 2000

S. Linko
Medix Diacor Labservice,
Department of Quality Assurance,
Nihtisillankuja 1, 00260 Espoo, Finland
e-mail: solveig.linko@medix.fi
Tel.: +358-9-5256-277
Fax: +358-9-5256-260

Abstract The Quality Assurance Department of Medix Diacor Labservice evaluated a two-way method validation procedure for serum lithium quantification in therapeutic drug monitoring In the process of a company fusion and rationalization of two considerably large production lines, three independent ion-selective electrode (ISE) methods were surveyed, among many others. While tailoring the new medical laboratory production, subcontracting from a collaborating company was discontinued. Likewise, modernization of the ISE instrumentation was unavoidable to increase throughput and effectiveness. It was important that the new result levels should be comparable both with the former subcontractor's levels and with the levels reported from the previously existing instrumentation. The aim of this study was to evaluate the most crucial performance characteristics of a novel lithium method in comparison to the two ISE test methods being withdrawn. The standardized lithium test method was inspected in terms of linear measurement range, analytical variation, bias, past and on-going proficiency testing, in addition to method comparison, to achieve the desired analytical goals. Fulfilling the accreditation requirements in terms of the introduced method validation parameters is discussed.

Keywords Validation · Lithium · Performance · Proficiency testing · Analytical goals

Introduction

Lithium is used as a therapeutic drug for patients suffering from manic-depressive illness and other mood disorders. Serum concentrations should be monitored regularly in all patients receiving lithium therapy. Therapeutic drug monitoring (TDM) and avoidance of intoxication are the key indications of serum lithium (S-Li) measurements. TDM is as important in the treatment of acute phase psychiatric disorders as in long-term therapy. The therapeutic concentration range is relatively narrow, 0.60 to 1.2 mmol/l [1]. Adverse reactions may occur at concentration levels of 1.0 to 1.5 mmol/l and more severe reactions at higher concentrations [2, 3]. Thus, analytical quality specifications must be established and well-known prior to routine production with any new measurement system. Accordingly, ensuring the trueness of result level and the appropriate analytical performance are fundamental points.

In 1990, Westgard et al. reported on the importance of understanding quality management science (QMS) in clinical chemistry [4]. Throughout the 1990s, many guidelines for implementing quality systems in medical laboratories and explanations on criteria required for medical laboratory accreditation were introduced [5–7]. Today quality management with its various elements should be part of the routine work in a modern clinical laboratory, whether the laboratory is accredited or not.

The fantastic development of technology in the medical laboratory branch, as in many others, allows unbreakable validation processes at many levels.

If new test methods combined with new instrumentation are introduced into a laboratory they should, self-evidently, be state of the art . On the other hand, if minor modifications occur in existing methods, obviously there will be less parameters to be validated. Therefore, the validation process should be defined and well-planned in all cases. The European co-operation for Accreditation of Laboratories (EAL) has provided general guidelines on certain issues related to the validation of test methods [8]. According to paragraph 2.1.3, the required depth of the validation process depends on the maturity of the test method and the prevalence of its use. "A standardised test method" is exclusively defined by EAL. This interpretation can easily be applied to the medical laboratory field and has been discussed [9] in terms of additionally needed clarification in the new ISO/IEC 17025 standard [10]. Paragraph 5.4.5 states that the validation shall be as extensive as is necessary to meet the needs of the given application or field of application.

Ion-selective electrodes (ISEs) represent the current primary methodology in the quantification of S-Li [11–13]. Moreover, ISE modules are parts of large and fully automated clinical chemistry analysers. In practice, the validation parameters are most often chosen in terms of judging the acceptability of the new measurement system for daily use. For this reason, the first approach was to study whether the detected imprecision fulfilled the desired analytical quality specifications. Secondly, proficiency testing (PT) results from past samples were of great value in predicting future bias. The identity of the three ISE methods was evaluated using patient samples. The analytical performance was checked after 6 months routine use. Without any exception, method validations always mean an extra economical burden. Therefore, the validation parameters chosen and processed have to be considered carefully.

The ISE method validation was performed in two directions (two-way), during the course of a laboratory fusion. The two laboratories involved in the fusion were both previously accredited according to EN 45001. Nevertheless, an S-Li test method was not included in either of the scopes of accreditation.

Background to the two-way method validation process

Originally three independent clinical laboratories were reporting S-Li results to clinicians, each using of their own ISE methodology. Laboratory A (Lab A) subcontracted the S-Li tests from its co-operating laboratory B (Lab B) for many years. Meanwhile, laboratory C (Lab C) produced and reported its state-of-the-art S-Li results independently of Lab A and Lab B. In 1999, Lab A and Lab C fused to form one company. In the new laboratory union, the former Lab A was lacking an in-house S-Li test method, while the current instrumentation used for S-Li determinations was running out of its technical performance capabilities. Consequently, New Lab A (united Lab A+C) evaluated the production status in terms of S-Li and decided to set up a fully automated S-Li method within the capabilities of the available clinical chemistry automation and discontinue the subcontract with Lab B.

Two problems needed to be resolved: 1) How to perform an adequate method validation to verify the S-Li concentrations reported by the new laboratory? and 2) Which validation steps would fulfil the required criteria, if the laboratory wanted to have the S-Li ISE method considered within its scope of accreditation in the future?

Experimental

Instrumentation

A Cobas Integra 700 clinical chemistry analyser combined with an ISE module from Roche Diagnostics (Germany) was used by the New Lab A. The subcontractor, Lab B tested S-Li using a Microlyte 6 ISE analyser from Kone Instruments (Finland). Laboratory C applied a 654 $Na^+/K^+/Li^+$ analyser (Chiron 654) from Ciba Corning Diagnostics Ltd (Chiron Instruments, UK) for its S-Li production.

Sample material

Sixty two serum samples for method comparison were collected from the patient sample pool of Lab A. The S-Li concentrations varied between 0.1 mmol/l and 1.6 mmol/l, thus covering both the therapeutic range and the decision making levels of S-Li. All 3 instruments were capable of measuring, quantitatively, 56 samples. Six samples, were beyond the lower measurement ranges of the Microlyte 6 and Chiron 654 instruments. The samples were aliquoted into appropriate tubes and refrigerated until analysed.

The PT material was obtained from the national scheme organizer, Labquality (Helsinki, Finland) and from an international scheme organizer, Murex Biotechnology Ltd. (UK). The samples were of human origin and purchased either as liquid or lyophilized material. All PT samples were prepared and stored until analysed according to the instructions given by the supplier.

Reagents

The three ISE setups each required their own system solutions and calibrators which were purchased from the respective manufacturers. A stock solution of 50 mM LiCl in deionized water was serially diluted in lithium-free fresh human serum to test the linearity of the Cobas Integra 700 ISE method. The tested concentration range was from 0.06 mmol/l to 4.01 mmol/l of lithium. LiCl, p.a. 99% purity, was purchased from Merck (Germany).

ISE methods

The ISE systems of all three analysers measured lithium in undiluted serum as direct measurements. Automated system calibration and calculation of the results were in-built in all the instruments. The lithium measurement range of the Cobas Integra 700 ISE module was from 0.10 mmol/l up to 4.00 mmol/l. According to the manufacturer's specifications the linear measurement range of the Kone Microlyte 6 ISE analyser was from 0.20 mmol/l to 4.00 mmol/l and for the Chiron 654 ISE from 0.20 mmol/l to 5.00 mmol/l.

A long-term quality control serum, Longtrol (Labquality, Finland), was used by Lab B and Lab C for daily internal quality control. New Lab A used Daytrol as well as normal and pathological quality control samples from Roche Diagnostics for both method validation and routine use. Two different lots of Longtrol and the Roche controls were used in this study.

Statistical analysis

All statistical tests were run using Analyse-It with Microsoft Excel 5.0 for Windows (Analyse-It Software Ltd., 40 Castle Ings Gardens, Leeds, UK). Ordinary linear regression (OLR) was used in testing the linear measurement range of the novel ISE method. The normality of the method comparison for the three ISE setups were checked by Shapiro-Wilk W test. Agreement between the ISE instrumentation was elucidated with Altman-Bland (AB) plots. Passing and Bablok regression analysis was used for method comparisons. The relative biases from PT results were calculated according to the Marchandise equation [14].

Results and discussion

Linearity

The mean values of the replicate measurements in eight serial dilutions were plotted against the theoretical lithium values. The method showed acceptable linearity between 0.10 and 4.00 mmol/l of lithium ($y=0.970x+0.014$, $R^2=0.9996$) and the results were comparable to the manufacturer's specifications. The 95% confidence intervals (CIs) for the intercept and slope are given in Fig. 1.

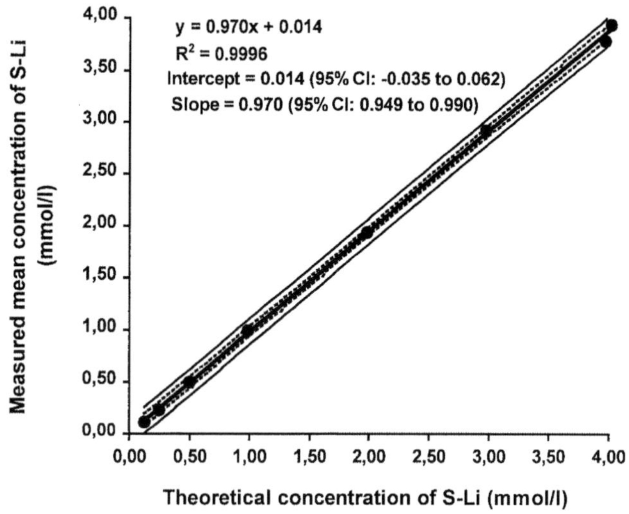

Fig. 1 The linearity of the Cobas Integra700 ion-selective electode (ISE) lithium method. S-Li serum lithium; linear regression line $y=0.970x+0.014$; $R^2=0.9996$; intercept $=0.014$ (95% CI: -0.035 to 0.062); slope $=0.970$ (95% CI: 0.949 to 0.990)

Imprecision

The analytical variation of the Cobas Integra ISE method in New Lab A was studied in the validation process by running the two-level system quality control samples for10 days, and Daytrol for8 days. The intra-assay variations (n=20) of the two system controls were much better than those quoted by the manufacturer. The within-batch variation of the lower system control was verified as 0.43% compared to 0.81% given by the manufacturer. The higher level varied by 1.4%: the manufacturer quoted a respective variation of 2.5%.

Lab B reported its inter-assay variation as total variation of Daytrol only at the higher therapeutic level. Respectively, Lab C reported variation only at the lower therapeutic level (Table 1). As the national target coefficient of variation (CV%) for analytical variation is 2% for S-Li, all methods fulfilled this quality specification, when the long-term control, Daytrol, was used. Inter-assay variation of the lower level system control was 3.5 times higher, than quoted by the manufacturer. This stresses the point that every laboratory should check the performance specifications given by the manufacturer. The variation of the higher therapeutic level showed equality with the obtained validation results.

Table 1 Inter-assay variation, CV_{meas} % of the three ion-selective electrode (ISE) applications during method validation and at the 6-month checkpoint for the lower and higher therapeutic levels and the toxic level of serum lithium (S-Li)

ISE application	CV_{meas} % and $(CV_{man})^a$ % (Lithium level)		
Cobas Integra 700 "Method validation"	1.7^b(0.48) (0.48 mmol/l)	2.4^c (1.14 mmol/l)	1.5^d (2.5) (2.11 mmol/l)
Cobas Integra 700 "Six-month checkpoint"	3.3^b (0.48) (0.53 mmol/l)	2.8^c (1.14 mmol/l)	4.5^d (2.5) (1.86 mmol/l)
Kone Microlyte 6 "Method validation"	–	1.8^c (1.14 mmol/l)	–
Chiron 654 "Method validation"	0.91^e (0.67 mmol/l)	–	–

[a] CV_{man} is the inter-assay variation specified by the manufacturer
[b] Cobas Integra 700 system control, normal level
[c] Daytrol, lot "a"
[d] Cobas Integra 700 system control, pathological level
[e] Daytrol, lot "b"

Table 2 Relative bias percentages of the three ISE methods. CP is 6-month checkpoint; QAP is Quality Assessment Programme

Relative bias %			
ISE instrumentation	Method validation National QAP	CP National QAP	CP International QAP
Cobas Integra 700	3.3 (n=6)	3.2 (n=6)	2.8 (n=12)
Kone Microlyte 6	5.7 (n=6)	–	–
Chiron 654	7.3 (n=6)	–	–

At the 6-month checkpoint the variations of both system controls showed much worse performance compared to the results measured during the validation process. Usually one expects better performance in variation over a 12 times longer time period. An explanation for the higher between-day variation might be differences between the two lots of the control samples and/or over-all unsuitability of this control material. These results demonstrate, that imprecision detected during any validation gives only an idea of the real variation. Further corrective actions should be done to check the high variation of the system controls. A temporary instability of the lithium electrode could not explain this variation. The cumulative report of Daytrol from the first 6 months showed satisfactory performance. Ninety six percent of the results were within the target of 2%, although the cumulative variation was 2.8%. Higher than allowed variation was detected during two summer months, which explains why the allowable target was exceeded. The proposed European specifications for imprecision of S-Li measurements is 3.6% at the decision levels of 0.4 mmol/l and 1.5 mmol/l [15, 16]. Excluding the higher system control at the six-month checkpoint, the variations of all other controls in every ISE system were within this target.

Bias

The national allowable total analytical error, TE_a is 6%. According to the European specifications for inaccuracy, the respective allowable bias is maximally 4.2% [15, 16]. Table 2 shows the relative biases calculated from the PT results according to the Marchandise equation. Six past samples from Labquality were run on the Cobas Integra ISE and Chiron 654 analysers in the validation process. The subcontractor, Lab B, reported their results from the respective months. All biases of the national PT samples were related to consensus means ($n=72\pm2$) within the method group of "liquid chemistry". None of these past samples had reference method values (RMVs). It can be concluded from the calculated relative bias that the novel method showed excellent performance. The Chiron 654 results were much more biased and beyond the acceptable limits of both the national and European quality specifications. Neither of the relative biases of the novel method exceeded the target limits either in national or international schemes at the 6-month checkpoint. Two of the PT samples during that time period had traceable values to flame emission photometry (INSTAND Reference Laboratories, Düsseldorf, Germany) as the reference method. Unfortunately, the uncertainties, from the two RMVs could not be estimated because the uncertainty, U_i of the reference method was unknown. On the other hand, the Marchandise equation can only be applied, if $n \geq 3$. The results in Table 3 show that although the consensus mean values (CMVs) did not deviate more than 2.4% (survey A) and 0.83% (survey B) from the RMVs, the new method was roughly within acceptable quality specifications in survey A, when the results was compared to the CMV. Anyhow, the performance was good in survey B and passed both national and European operation criteria.

Table 3 Comparison of bias percentages from reference method values (RMV) and consensus mean values (CMVs) within two national short-term (ST) surveys

National QAP	Bias % (95% CI) from RMV	Bias % from CMV
ST survey A	8.7 (0.957 mmol/l)	6.1 (0.98 mmol/l)
ST survey B	3.3 (2.40 mmol/l)	2.5 (2.42 mmol/l)

Method comparison

Fifty six of the sixty two samples gave a quantitative lithium concentration with all three ISEs. Six samples were beyond the lower measurement level (0.20 mmol/l) of the Kone Microlyte 6 and Chiron 654 analysers. As these results were not comparable to the results obtained by the novel method, they were excluded from the statistical analysis and further discussions. The probability values (*P*) of the Shapiro-Wilk W coefficients were 0.07 for the Cobas Integra instrument, 0.04 for the Kone Microlyte 6 and 0.09 for the Chiron 654 as the normalities of the result groups were tested. In the descriptive comparison of the three ISEs it was found that the new method and the Kone Microlyte 6 method gave comparable results, while the result level of the Chiron method was approximately 10% higher than those of the other two (Table 4). In addition to the skewness detected in all result groups, it was concluded that none of them were normally distributed. The Altman-Bland (AB) agreements between the two comparison ISEs and the new method are shown in Figs. 2 and 3. A positive bias of 0.013 mmol/l (95% CI: 0.005 to 0.021) was detected between the Cobas Integra and Kone Microlyte 6 methods. This bias was not of clinical significance and only 3.6% of the 56 samples were beyond the upper limit of 95% of agreement. The Cobas Integra method was negatively biased compared to the Chiron 654. From Fig. 3 it can be seen that three samples (5.4%) fell outside the 95% limits of agreement. The bias of –0.052 mmol/l (95% CI: –0.059 to –0.045) was obvious due to the largest relative bias of PT samples (Table 2) despite the best imprecision of the Chiron instrument (Table 1). According to the physicians' opinion this deviation was not of great clinical significance. Unfortunately, neither of the ISE methods being withdrawn could be compared to the RMVs to see the biases from the true values.

Passing and Bablok regression analysis between the methods resulted in biases as predicted in AB agreement (Table 5). The CIs of the intercepts and the slopes did not overlap between the comparison instruments. It can be concluded that although a statistical difference was detected, this was not clinically significant either. Lab A previously reported a therapeutic range of 0.8–1.2 mmol/l, and Lab C a range of 0.5–1.5 mmol/l with their patient results. Thus, accord-

Table 4 Descriptive comparison of the three ISE instruments

ISE instrumentation	n	Median S-Li (mmol/l)	95% CI of median S-Li (mmol/l)
Cobas Integra 700	56	0.640	0.540 to 0.680
Kone Microlyte 6	56	0.625	0.520 to 0.660
Chiron 654	56	0.700	0.610 to 0.740

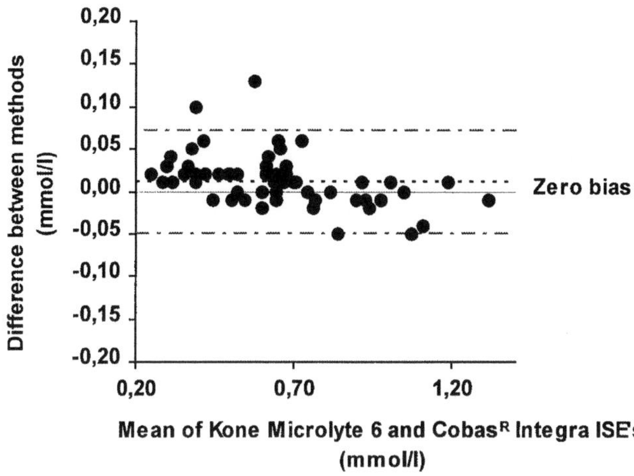

Fig. 2 Agreement between the Cobas Integra 700 and Kone Microlyte 6 methods as an Altman-Bland (AB) plot. The 95% limits of agreement are illustrated as dashed lines

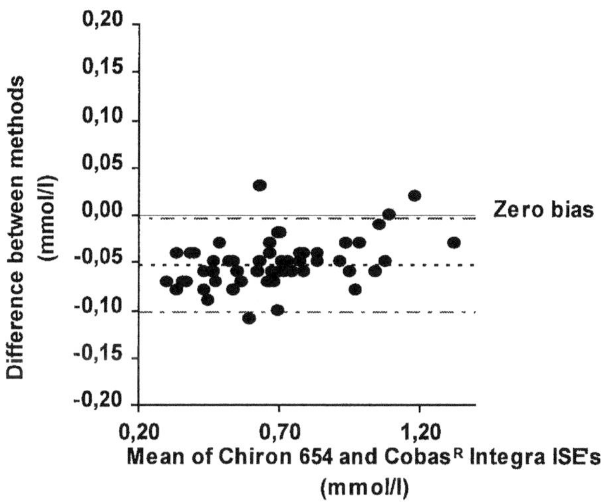

Fig. 3 Agreement between the Cobas Integra 700 and Chiron 654 methods as an Altman-Bland (AB) plot. The 95% limits of agreement are illustrated as dashed lines

Table 5 The summarized data of the ISE method comparison (n=56)

Cobas Integra 700 *v.* the comparison methods	AB agreement	Passing and Bablok method conversion	
	Bias	Intercept	Slope
Kone Microlyte 6	0.013 (95% CI: 0.005 to 0.021)	0.041[a]	0.951[b]
Chiron 654	−0.052 (95% CI: −0.059 to −0.045)	−0.084[c]	1.041[d]

[a] Kone Microlyte 6:95% CI of intercept 0.026 to 0.057
[b] Kone Microlyte 6:95% CI of slope 0.925 to 0.977
[c] Chiron 654:95% CI of intercept –0.099 to –0.055
[d] Chiron 654:95% CI of slope 1.000 to 1.067

ing to these results a therapeutic range of 0.6–1.2 mmol/l and a toxic range of >1.6 mmol/l could reliably be established at New Lab A.

Conclusions

This method validation study indicates that the choice of validation parameters is crucial. Checking the method performance in accordance to the obtained validation results should not be omitted. The preliminary imprecision results do not necessarily predict the analytical variation as the method is taken into routine use. The suitability of system controls have to be considered carefully when patient results are accepted according to the general internal quality control principles. Further, surprisingly high variations should be clarified with the manufacturer. The use of past PT samples is highly recommended to predict the future inaccuracy, whenever possible. However, it would be of great value to estimate the true bias from RMVs with given uncertainties and more frequent availability. The deviations from the RMVs could consequently be used in uncertainty budgeting and the estimation of the total measurement uncertainty. Certified reference materials with traceable values cannot be used in single method validations due to their high expense.

It was concluded that the new ISE method presented for the determination of S-Li fulfilled the analytical quality specifications. The results from the 6-month checkpoint were satisfactory and could be used for further assessment by the national accreditation body in the future.

Acknowledgements The author wishes to give many thanks to Aino Salla for her technical assistance and to chief chemist Paavo Tammisto at Rinnekoti Foundation Laboratory (Espoo, Finland) for his many contributions to this study.

References

1. Tietz NW (1995) Clinical guide to laboratory tests, 3rd edn. W.B. Saunders, USA, p. 840
2. Price LH, Heninger GR (1994) N Engl J Med 331:591–597
3. NIMH/NIH Consensus Development Conference Statement (1985) Am J Psychiatry 142:4
4. Westgard JO, Burnett RW, Bowers GN (1990) Clin Chem 36:1712–1716
5. Dybkær R, Jordal R, Jørgensen PJ, Hanson P, Hjelm M, Kaihola HL, Kallner A, Rustad P, Uldall A, de Verdier CH (1993) Scand J Clin Lab Invest 53 Suppl. 212:60–84
6. European Communities Confederation of Clinical Chemistry (ECCCC) (1996) Essential criteria for quality systems of clinical biochemistry laboratories. In the EU Working Group on Harmonization of Quality Systems and Accreditation, EC4
7. European cooperation for Accreditation of Laboratories (EAL) (1997) Accreditation for medical laboratories, EAL-G25/ECLM-1, 1st edn. European cooperation for Accreditation (EA)
8. European cooperation for Accreditation of Laboratories (EAL) (1997) Validation of test methods. General principles and concepts, EAL-P11, 1st edn. European cooperation for Accreditation (EA)
9. Walsh MC (1999) Accred Qual Assur 4:365–368
10. ISO/IEC 17025 (1999) General requirements for the competence of testing and calibration laboratories. International Organisation for Standardisation (ISO), Geneva, Switzerland
11. Linder MW, Keck PE Jr. (1998) Clin Chem 44:1073–1084
12. Sampson M, Ruddel M, Elin R (1994) Clin Chem 40:869–872
13. Kuhn T, Blum R, Hildner HP, Wahl HP. (1994) Clin Chem 40:1059
14. Marchandise H (1993) J Anal Chem 34:82–86
15. Fraser CG, Petersen HP, Ricos C, Haeckel R (1992) Eur J Clin Chem Clin Biochem 30:311–317
16. Westgard JO, Seehafer JJ, Barry PL (1994) Clin Chem 40:1228–1232

Accred Qual Assur (2001) 6:173–177

Peter Woitke
Reinhard Kreßner
Peter Lepom

Determination of hydrocarbons in water – interlaboratory method validation before routine monitoring

Received: 31 October 2000
Accepted: 3 January 2001

Presented at the EURACHEM/EQUALM Workshop "Proficiency Testing in Analytical Chemistry, Microbiology and Laboratory Medicine", 24–26 September 2000, Borås, Sweden

P. Woitke (✉) · R. Kreßner · P. Lepom
Federal Environmental Agency,
P.O. Box 33 00 22, 14191 Berlin, Germany
e-mail: peter.woitke@uba.de
Tel.: +49 30 8903 2566
Fax: +49 30 8903 2285

Abstract The clarification of hydrocarbon input into the Baltic sea via rivers is one of the priority issues of the 4th Pollution Load Compilation (PLC-4) within the framework of international Baltic Sea marine monitoring. An interlaboratory comparison was conducted to check the applicability of a new method for the determination of hydrocarbons by solvent extraction and gas chromatography. Surrogate oil solutions with known hydrocarbon content were distributed among the participants for preparation of water samples of different hydrocarbon concentration. In using these concentrations as assigned values and by setting target values for precision, the proficiency of participating laboratories could be tested as a qualifying step before involvement in PLC-4 routine monitoring. The results of the exercise indicate that hydrocarbons in water samples can be monitored as a mandatory test item within the framework of PLC-4.

Keywords Hydrocarbons · Mineral oils · Water monitoring · Quality assurance · Interlaboratory comparison

Introduction

Nowadays it is generally accepted that laboratory proficiency testing schemes are an important quality assurance tool in environmental monitoring programmes. Apart from using validated analytical methods and internal laboratory quality control procedures, the regular participation of laboratories in interlaboratory comparisons is required to ensure the accuracy and comparability of data.

Within the framework of the 4th Pollution Load Compilation (PLC-4) of Baltic Sea monitoring, the clarification of oil inputs to the Baltic Sea via rivers is one of the priority issues according to Helsinki Commission (HELCOM) recommendations [1]. Hence, the determination of hydrocarbons in some larger rivers and point sources is mandatory in PLC-4. Gas chromatographic (GC) determination of hydrocarbons after solvent extraction was chosen as the analytical procedure [2]. The method enables the determination of hydrocarbons at concentrations above 0.1 mg L^{-1}, and encompasses a wide range of aliphatic, alicyclic, and aromatic compounds of interest. Compared with the frequently applied infrared spectrometric method, the GC procedure provides much additional information relating to boiling point range and qualitative composition of the hydrocarbon contamination.

Oil inputs are to be measured within PLC-4 for the first time. To support external quality assurance measures and to obtain information on the proficiency of the nominated laboratories, an interlaboratory comparison on the determination of hydrocarbons in water was conducted before the start of routine monitoring.

Test materials

Because hydrocarbons, especially *n*-alkanes, in aqueous solution are susceptible to microbiological degradation, surrogate samples were used in the interlaboratory comparison. All participants were requested to prepare water samples from the solutions listed in Table 1 according to

Table 1 Solutions for preparing surrogate samples distributed in the interlaboratory comparison on the determination of hydrocarbons in water

Sample	Code	Concentration	Solvent	Volume [mL]	Weight of hydrocarbons [mg]		Weight of interfering compounds [mg]		
					Diesel fuel [mg]	Lubricating oil [mg]	Texapon NSO (surfactant)	Dodecanoic acid methyl ester	River fulvic acid
Standard solution	STD	0.624 mg mL^{-1}	*n*-Hexane	50	20.2	11	–	–	–
Surrogate sample	S1	0.224 mg L^{-1}*	*i*-Propanol	250	33.9	16.4	–	–	–
Surrogate sample	S2	2.22 mg L^{-1}*	*i*-Propanol	250	247.8	252.3	–	–	–
Surrogate sample	S3	0.474 mg L^{-1}*	*i*-Propanol	250	72.2	34.4	253	255.6	41.2
Surrogate sample	S4	3.34 mg L^{-1}*	*i*-Propanol	250	381.4	369.8	250.1	249.8	41.7

*If prepared according to the sample preparation instructions

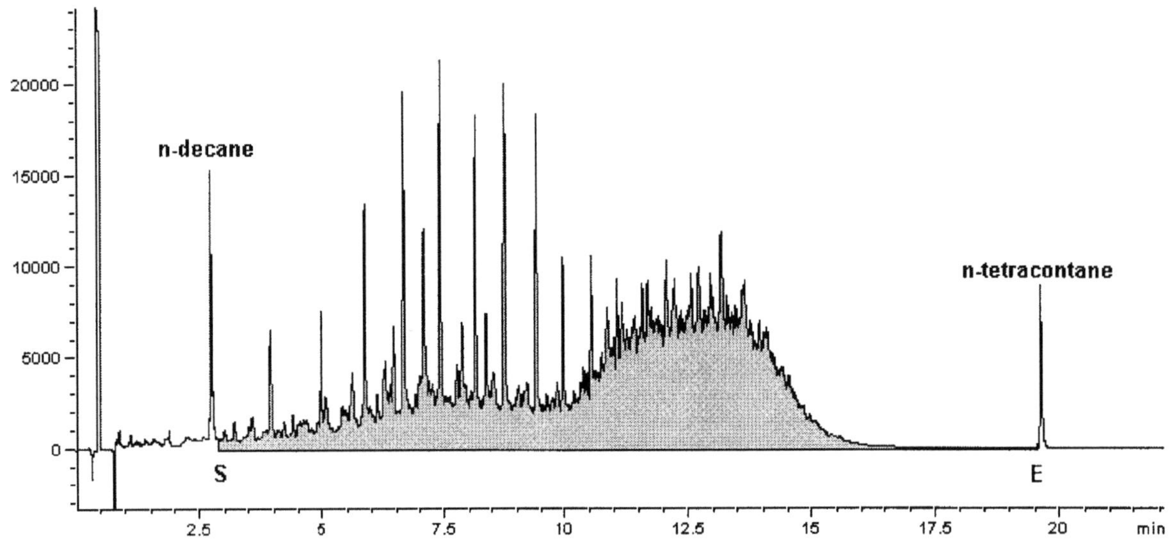

Fig. 1 Integrated gas chromatogram of a hydrocarbon standard solution corrected for 'column bleed' (1 mg mL^{-1} in *n*-hexane, 3 μL on-column injection, 12 m × 0.32 mm BPX-5 column, 60 °C for 1 min, 20 °C/min to 360 °C, 360 °C for 15 min)

instructions distributed with the samples. All samples were weighed before shipment to enable monitoring of solvent evaporation.

Two concentration levels were to be analysed by the participants, one near the limit of determination and the other ca. ten times this level. The analytical procedure includes a clean-up step in which polar compounds, e.g. plant fats, fulvic acids, and surfactants, are removed from the extracts. To check the influence on analytical results of typical matrix constituents which might be present in river waters and waste waters, possible interfering compounds were added to a subset of the test materials (Table 1).

Because the hydrocarbon test samples were provided as solutions and were prepared from only one stock solution, a test of homogeneity was not necessary. A subset of sample aliquots was kept by the organiser for stability testing. The sample weight was recorded and was found to be constant during the time of the exercise.

Interlaboratory comparison

The hydrocarbon oil index within PLC-4 was determined in accordance with the draft standard ISO/DIS 9377–4, which has recently been adopted as ISO/FDIS 9377–2 [2]. Briefly, water samples are extracted with *n*-hexane or petroleum ether. After clean-up of the extracts on a Florisil column, hydrocarbons are detected and quantified by capillary gas chromatography with flame ionisation detection (GC-FID, Fig. 1).

Results from previous interlaboratory validation studies on the method showed accuracy to be acceptable – relative reproducibility standard deviations were between 20% and 40% depending on the hydrocarbon concentration and the amount of interfering compounds in the sample. Recoveries between 80% and 100% were obtained for most samples. In the presence of surfactants, only, recoveries dropped to 60%.

Because the number of PLC-4 laboratories officially nominated by the HELCOM countries was very limited, additional laboratories were invited to participate. Laboratories not familiar with the new method received several test samples to acquaint them with the procedure before announcement of the exercise. Samples, data report sheets, and a questionnaire referring to the experimental conditions were sent to 24 laboratories. The participants were requested to perform four replicate determinations. Sixteen laboratories in six different countries around the Baltic Sea sent in results to the organiser.

The data were assessed by the ISO 5725–2 protocol [3] implemented in the software package Prolab98 (Dr Uhlig, Quo Data, Dresden, Germany), which is routinely used by the German Federal Environmental Agency for evaluation of laboratory proficiency tests. Outliers were rejected by use of Grubbs' test (P=1%) and Cochran's test (P=1%).

Results

Because samples S1–S4 were synthetic solutions of mineral oils, the overall means of the participants can be compared with assigned values, and recoveries can be calculated (Fig. 2). For sample S1 the recovery was in the expected range above 85%. It was only 75% for sample S2 indicating incomplete extraction of higher concentrations of hydrocarbons. Recovery of ca. 70% was obtained for samples S3 and S4 which contained interfering compounds, irrespective of mineral oil concentration. To check the calibration functions established by the participants, a standard solution (STD) containing an unknown concentration of mineral oil for direct injection into the GC system was distributed. Evaluation of the results revealed that the correct concentration was found (Fig. 2), although the relatively high overall repeatability and reproducibility obtained for the standard solution must be regarded as insufficient.

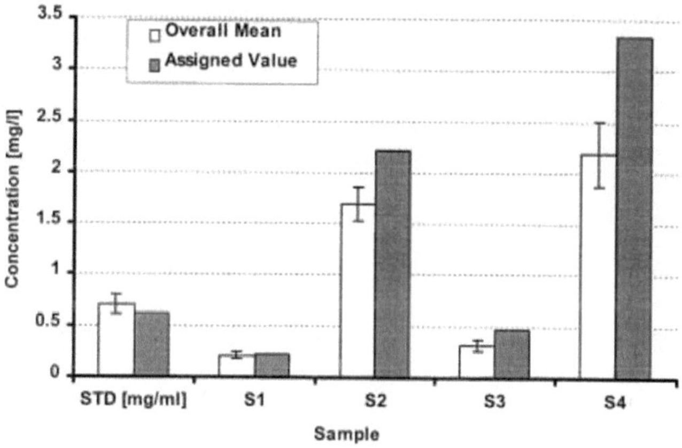

Fig. 2 Comparison of assigned values and participants' overall mean obtained for the standard solution, STD, and water samples S1–S4 (error bars indicate the 95% confidence interval)

Table 2 summarises the overall statistics of the interlaboratory comparison. Repeatability between 7% and 16% and reproducibility of ca. 30% were obtained for all the samples investigated, in good agreement with results from a recently conducted interlaboratory validation study arranged by ISO/TC147/SC2/WG15 [2]. The results indicate that mineral oil concentrations in water samples between 0.2 and 3 mg L^{-1} can be measured with adequate precision by the participating laboratories.

There were no significant differences between the repeatability or the reproducibility obtained for the standard solution and those obtained for the different water samples. Thus the variability of the sample preparation steps did not seem to contribute markedly to the overall uncertainty of the analytical procedure. It could be concluded that variability in the determination of hydrocarbons in water is mainly a result of the precision of the GC determination.

Proficiency evaluation

Analysis of systematic errors in the determination of hydrocarbons in water can be achieved by use of Youden plots after transformation of the results to an overall mean of M_{total}=0 and a standard deviation of S_{total}=1 (Fig. 3). Almost all laboratories are distributed around the 45° line indicating that most of the variation was systematic rather than random, particularly at higher mineral oil concentrations (sample pair S2/S4, Fig. 3B). Results located within the interval $M_{total}\pm 2\ S_{total}$ indicate sufficient proficiency of the participating laboratories in performing the determination of hydrocarbons in water

Table 2 Summary statistics of the HELCOM interlaboratory comparison exercise on the determination of oil in water

Sample	No. of labs	No. of results	No. of outliers	Overall mean [mg L^{-1}]	Reference value [mg L^{-1}]	Overall recovery [%]	S_r* [%]	S_R* [%]
STD	14	49	4	0.717**	0.624**	115	7.45	26.4
S1	14	50	8	0.206	0.224	92.0	14.8	31.3
S2	15	56	4	1.69	2.22	76.1	9.24	21.9
S3	14	46	4	0.320	0.474	67.5	16.2	34.9
S4	15	58	4	2.20	3.34	65.9	9.58	30.2

* S_r – relative repeatability standard deviation; S_R – relative reproducibility standard deviation
** [mg mL^{-1}]

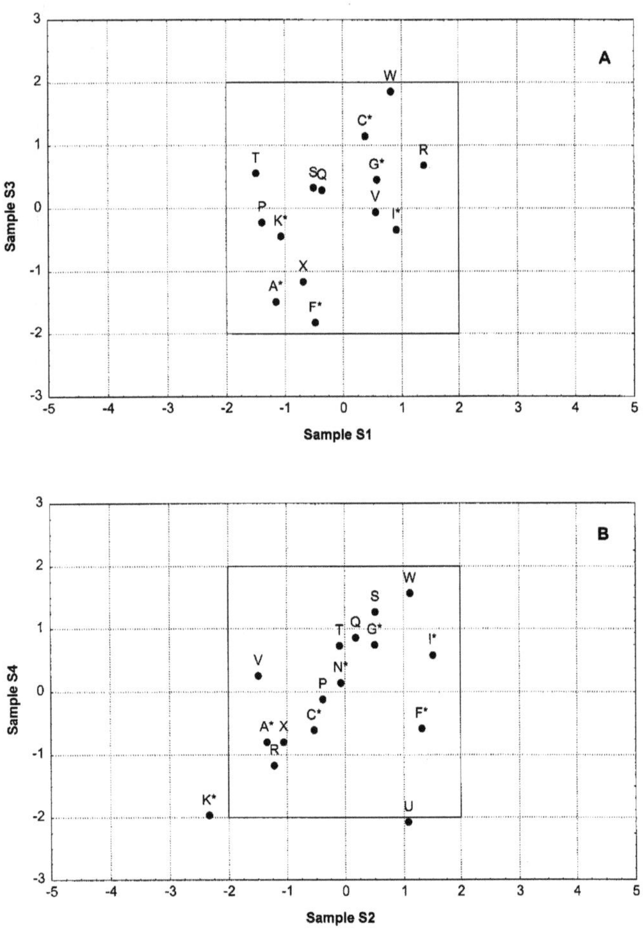

Fig. 3 Youden plot of the results (transformed to an overall mean of M_{total}=0 and a standard deviation of S_{total}=1) for sample pair S1/S3 (**A**) and sample pair S2/S4 (**B**). Satisfactory performance is represented by the displayed box (M_{total}±2 S_{total}). Each dot represents the results of an individual laboratory marked by a letter code (* – laboratories involved in PLC-4)

samples. For comparison purposes results from another laboratory (participant K), which used a different method of determination (fluorescence spectroscopy) are included. For higher mineral oil concentrations, this method seems to have a negative systematic error.

The calculation of the Z-scores to evaluate the laboratory's proficiency [4] was based on the known concentrations of hydrocarbons in the samples (Table 2) and on a target precision set by the PLC-4 programme:

20% for hydrocarbon concentrations above 1 mg L^{-1}; and

30% for hydrocarbon concentrations below 1 mg L^{-1} (near the determination limit).

The Z-scores achieved by the participants are displayed in Fig. 4 for all the samples investigated. Eleven laboratories completed the interlaboratory comparison successfully (80% of Z-scores within |Z|< 2). The between-laboratory standard deviations are better than the target values for tolerable error set by PLC-4 (Table 2). Unsatisfactory Z-scores of individual laboratories were mainly a consequence of low recoveries, in particular for samples containing interfering substances. As a consequence, correction of analytical results for recovery, regularly determined by analysis of spiked water samples, will be conducted within the PLC-4 programme.

Summarising the results of this study, mineral oils can be determined in water samples by solvent extraction and gas chromatography. This test item will be mandatory within the framework of PLC-4. Nevertheless, further interlaboratory comparisons should be performed in the course of PLC-4 as an external quality-assurance measure within this monitoring programme.

Acknowledgements The authors would like to thank all laboratories for participating in the interlaboratory comparison. Financial support from HELCOM is gratefully acknowledged.

Fig. 4 Z-scores of the participating laboratories (A–X) for samples S1–S4 calculated using assigned concentrations and PLC-4 target values for precision (the critical Z-value |Z|=2 is plotted as a *bold line*, * – laboratories involved in PLC-4)

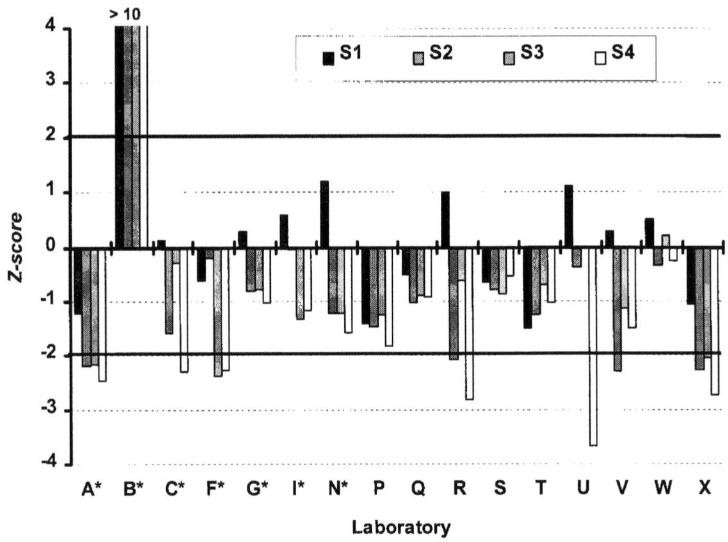

References

1. Helsinki Commission, Baltic Marine Environment Protection Commission (1999) Guidelines for the Fourth Baltic Sea Pollution Load Compilation (PLC-4)
2. ISO/FDIS 9377–2 (2000) Water Quality – Determination of Hydrocarbon Oil Index – Part 2: Method Using Solvent Extraction and Gas Chromatography
3. ISO 5725 (1994) Accuracy (Trueness and Precision) of Measurement Methods and Results – Part 2: Basic Method for the Determination of Repeatability and Reproducibility of a Standard Measurement Method
4. ISO/IEC Guide 43–1 (1997) Proficiency Testing by Interlaboratory Comparisons – Part 1: Development and Operation of Proficiency Testing Schemes

Accred Qual Assur (2003) 8:21–24
DOI 10.1007/s00769-002-0549-9

G. Anand Kumar
H.S. Sahoo
S.K. Satpathy
G. Swarnabala

Interlaboratory quality audit program for potable water – assessment of method validation done on inductively coupled plasma atomic emission spectrometer (ICP-AES)

Received: 3 April 2002
Accepted: 4 September 2002

G. Anand Kumar · H.S. Sahoo
S.K. Satpathy · G. Swarnabala (✉)
VIMTA Labs Ltd,
Hyderabad – 500 051, India
e-mail: swarnabalag@yahoo.co.in
Tel.: +91-40-7264141
Fax: +91-40-7263657

Abstract In an effort to assess the method validation done using ICP-AES in our laboratory for potable water, an Environmental Laboratory Approval Program organized by New York State Department of Health, Wadsworth Center providing the reference material has been undertaken for 14 trace elements and seven other chemical constituents. The certified means for the reference material and the results obtained in our laboratory are compared. The comparisons helped us assess the quality of our work. All the data from the inductively coupled plasma atomic emission spectrometer (ICP-AES) fall into the ranges specified. These data are intended to depict the quality of chemical analysis being conducted in our laboratory and to increase the level of confidence of our clientele in accepting our test reports. It should be further noted that while the technique may not be new, the model is new and the simultaneous detection of elements required validation for those of our clientele who are only familiar with sequential AAS and AES.

Keywords ICP-AES · Environmental Laboratory Approval Program – New York Department of Health, Wadsworth Center · Trace elements · Potable water

Introduction

Certain major, minor and trace elements tend to accumulate causing toxicity in potable water. The quantitative measurement of these elements helps in diagnosis and treatment of a range of disorders. Proficiency testing materials, when studied by an analytical method, help in method improvement and validation and lead to better results, increasing the confidence of the analysts in reporting the data to the customers. Due to the complex nature of the material in question, its variable compositions of the elements, the need to validate the method of analysis is mandatory on the part of the analyst. So the analysis can be authenticated with the help of studies on proficiency testing materials, thus increasing the confidence level for values reported.

In an effort to meet this demand, we have procured proficiency testing material of potable water from the Environmental Laboratory Approval Program being conducted by New York State Department of Health, Wadsworth Center. The chief of this institute has initiated the collaborative analysis program for analyzing the reference material for subsequent certification, with respect to major, minor and trace element concentrations. The whole study is used to analyze the waters of all kinds [1–6] such as drinking water – IS: 10500/1991, packaged natural drinking water – IS: 13428/1998, packaged drinking water – IS: 14543/1998, and water for quality tolerances for processed food industry – IS: 4251/1967. The characterization of these materials includes different sample preparation methods and analytical techniques being used by all the laboratories worldwide. Our laboratory is participating from India with NABL accreditation. We have determined 14 constituents by ICP-AES and seven by classical methods for the present study. The aim of this paper is to highlight the potential of ICP-AES in the generation of accurate analytical data for several trace elements in potable water samples and increase the confidence level of the customers who send us these

Table 1 Validation data for potable water reference materials – Environmental Laboratory Approval Program analyzed for the following elements by ICP-AES

Parameter (μg/L)	Result	APHA method	Study mean	Accept. limits	Score
Ag	87.0	3120	82.2	73.8–90.5	Sat
As	57.3	3120	51.1	44.8–57.3	Sat
Ba	777	3120	817.0	696.0–942.0	Sat
Be	8.74	3120	7.42	6.6–8.94	Sat
Cd	10.0	3120	8.59	6.9–10.4	Sat
Cr	99.0	3120	88.4	75.1–102.0	Sat
Cu	300	3120	328.0	293.0–359.0	Sat
Fe	541	3120	523.0	478.0–568.0	Sat
Mn	70.5	3120	74.20	68.50–79.80	Sat
Ni	219	3120	240.0	201.0–273.0	Sat
Pb	31	3120	33.80	23.60–43.80	Sat
Sb	45.95	3120	39.30	27.40–51.0	Sat
Se	130.0	3120	48.50	37.80–56.60	Unsat
Zn	479	3120	479.0	439.0–518.0	Sat

Sat, Satisfactory; Unsat, unsatisfactory.

Table 2 Validation data for potable water reference materials – Environmental Laboratory Approval Program analyzed for the following chemical parameters on ICP-AES

Parameter (mg/L)	Result	APHA method	Study mean	Accept. limits	Score
Na	12.6	3500	12.50	11.30–13.70	Sat
Nitrite	1.52	4500	1.51	1.28–1.74	Sat
Chloride	62.9	4500	59.10	55.10–63.10	Sat
Fluoride	5.1	4500	5.46	4.94–6.04	Sat
NO_3 as N	7.94	4500	7.91	7.19–8.79	Sat
Sulfate	44.4	4500	48.4	43.4–53.4	Sat
TDS	228	2540	244.0	162.0–326.0	Sat

TDS, Total dissolved salts; Sat, satisfactory.

samples. Since the programme had all the constituents determined, the data for the classical methods is also presented for completeness.

Experimental

Instrumentation

The ICP-AES used is a Varian-Radial Vista (Varian Instruments, Australia) associated with a microcomputer, software operating parameters, etc. as given in reference [7].

Materials

The water sample is obtained from the Environmental Laboratory Approval Program. We have used ICP-AES multi-element Reference Standard supplied by E. Merck with Lot. No: 0C030033 for standard calibration. These standards have certified data obtained from NIST as third-generation traceability, assuring reasonable accuracy in the estimations (Certificate of Analysis, Certipur – Reference Material; 11355 ICP multielement standad IV made from NIST standards reference materials, Lot No. 0C030033. Dr. Harald Untenecker, Central Analytical Laboratory, Merck). An echellogram showing all the elements is attached.

Sample procedure

To minimize the possibility of change in concentration, the ampoule is opened just before the analysis. The sample is prepared as follows: the ampoule temperature is adjusted to 20 °C prior to analysis. Approximately 900 mL of reagent water is added to a 1-L volumetric flask. Reagent grade nitric acid (5.0 mL) is added to the volumetric flask. The flask is swirled to mix. Ampoule ID no. is recorded. The ampoule is broken and 10 mL of the sample is transferred to the flask using a pipette. Then the flask is filled to mark with Milli-Q water and thoroughly shaken. The 14 elements are measured and the data are tabulated in Table 1.

Standard preparation

Aqueous standards are prepared from 1000 ppm E. Merck standard reference material supplied with NIST traceability. The standards are made up using 18 MΩ Milli-Q water with 1% v/v high quality nitric acid 65%-Fluka Chemie.

The following calibration standards for a five-point calibration curve have been made keeping in view the need to cover the range of 10, 2, 1, 0.1, and 0.01-ppm concentrations:

Std. 1: Ag, Ba, Cd, Cr, Cu, Fe, Mn, Ni, Pb, and Zn.
Std. 2: Be.
Std. 3: As, Se, and Sb.

Rinse and calibration blank solutions were prepared from 18 MΩ Milli-Q water with 5% HNO_3 as per the instructions provided by the approval program.

Methods used for chemical analysis

All the chemical analysis procedures are adopted from APHA methods [8]. These are specified in Table 2 along with the analyte.

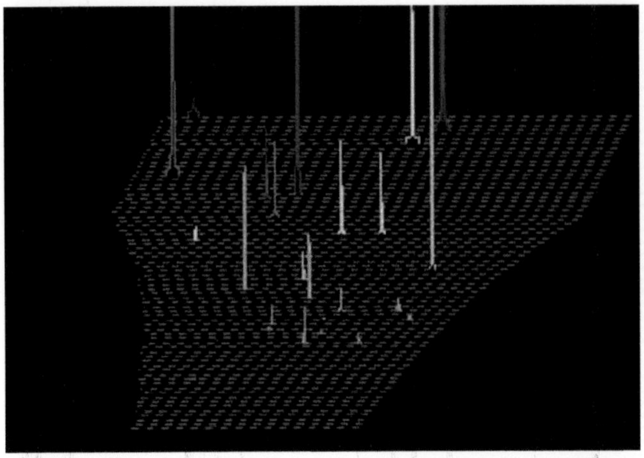

Fig. 1 Echellogram showing all the 23 elements in the standard along with the wavelengths. The wavelengths for all the 23 elements are:

Ag	328.068	Al	396.152	B	249.772	Ba	455.403
Bi	223.061	Ca	317.933	Cd	214.439	Cd	226.502
Co	238.892	Cr	267.716	Cr	357.868	Cu	327.395
Fe	238.204	Ga	294.363	In	230.606	K	766.491
Li	670.783	Mg	279.553	Mn	257.610	Na	589.592
Ni	231.604	Pb	220.353	Sr	407.771	Tl	190.794
Zn	213.857						

Results and discussion

ICP-AES offers rapid, multi-element determinations. Its sensitivity is lower than that of either ICP-MS or AA-GTA, but it can handle higher levels of total dissolved solids than ICP-MS and is much faster than AA-GTA. The results obtained are in comparison with all the equipment used by about 100 laboratories that are participating in the Environmental Laboratory Approval Program. The quality of the performance relating to equipment as well as the analyst in such international collaborative programs implies the use of certified reference materials for calibration and control samples to ensure a certain level of traceability of the measurements and finally the degree of assurance of reliability.

The use of reference standards for calibration minimized several analytical errors. However, our working standards also gave results very similar to the reference standards. The results obtained for the study in question are summarized in Tables 1 and 2. The organizers collected the assigned value for individual elements and provided us the mean or range of the submitted data for over 100 laboratories using highly reliable statistical procedures. The table gives the result, mean, method, range and reliability of the parameter in question. A careful observation of these data provides the precision with which the experimental work has been conducted in the present study. Figure 1 shows the Echellogram of 23 elements in the reference standard along with the wavelengths.

A typical example of the requirements for IS: 14543/1998 (4) are Ba, 1.0; Cu, 0.05; Fe, 0.1; Mn, 0.1; Zn, 5.0; Ag, 0.01; Al, 0.03; Se, 0.01; Ca, 75.0; Mg, 30.0; Na, 200.0; Hg, 0.001; Cd, 0.01; As, 0.05; Pb, 0.01; Cr, 0.05; Ni, 0.02; Sb, 0.005; B, 5.0 μg/L separately. Routinely all the samples are checked for the requirements before reporting. VISTA software has a built in program with USEPA requirements [9, 10] for the aspirating of standard reference solution; instrument check standard solution; interference check standard solution and QC standard solution for the laboratory QC control. Continuing calibration verification is done every ten analytical samples run on Auto Sampler, a facility that is not available when using the manual runs of the samples. We have adopted the Varian instruments instructions to calibrate the equipment to find the instrument detection limits and the method detection limits. The former values were obtained by aspirating on consecutive three days for all the standards and the latter values were obtained for consecutive seven runs on the same day for all the standards. These are also reconfirmed using "CHEMSW" software program available for the calculation of detection limits. The original "K-style" glass concentric nebulizer is used to obtain the raw-data with best possible sensitivity. For our method validation, we have adopted a seven-point calibration range for almost all the elements ranging between 10–0.01 mg/L. The accuracy in the recovery of the standards ranged from 95%–105% with a precision at 1 μg/L as ±12%. Matrix spiking results in the recoveries range between 75%–%125% as per USEPA/APHA. Vapour-generation accessory (VGA 77), is used for Hg measurements. Three standards 0.001; 0.005, and 0.01 μg/L are used for linearity or calibration. Stannous chloride (25%) and HCl (Fluka) are being used for the reduction in the cold-vapour generation.

Conclusions

It may be concluded that the observed data for 13 elements (except for the selenium) and the seven other constituents specified for potable water are in concurrence with all the laboratories whose data is submitted to Environmental Laboratory Program, New York State Department of Health. [After the constructive criticism from the reviewers, the unsatisfactory results for the Ag, Cd, Sb and Se elements were repeated. Cd values have improved by changing the wavelength from 217 to 226 nm. For antimony we have used a new standard and the value improved. For silver the number of washings before aspiration was tripled and the result improved. The selenium range is still not very satisfactory, but our efforts to improve upon this are in progress and we are positive about it.] The method detection limits, RSD's, and the instrumental detection limits obtained for our instrument and also the method validation done using this data confirms the validity of the method being adopted. Thus the use of instruments simultaneously for routine analysis imparts a high degree of excellence to the data being reported.

Acknowledgements G.S. is grateful to Dr. S. P. Vasireddi, Founder, Chairman and Managing Director, VIMTA Labs Ltd for giving permission to publish this work. We would like to thank the Environmental Laboratory Approval Program organized by New York State Department of Health, Wadsworth Center for permitting us to publish these data.

References

1. APHA (2000) Standard methods for the examination of water and waste water analysis, 20th edn. APHA.
2. (1991) IS 10500 – Drinking water
3. (1998) IS 13428 – Packaged natural mineral water
4. (1998) IS 14543 – Packaged drinking water (other than the packaged natural mineral water)
5. (1992) IS 1070 – Reagent-grade water
6. (1967) IS 4251 – Quality tolerances from water for processed food industry
7. Satyanarayana KVS, Ravi Kumar Ch, Swarnabala G (2002) Atomic Spectrosc (submitted)
8. APHA (2000). Methods – 3120, 3500, 4500 and 2540. In: Standard methods for the examination of water and waste water analysis, 20th edn. APHA
9. USEPA (1994) Determination of metal and trace elements in water and wastes by inductively coupled plasma–atomic emission Sspectrometry – Method 200.7, USEPA, Environmental Monitoring and Support Laboratory, Cincinnati, OH, May 1994. USEPA Washington, DC
10. Bridger S, Knowles M (2000) ICP-AES at work – ICP-29. A complete method for environmental samples by simultaneous axially viewed ICP-AES following USEPA guidelines. Varian Australia Pty Ltd.

Accred Qual Assur (2003) 8:108–112
DOI 10.1007/s00769-003-0598-8

Nineta Majcen

A need for clearer terminology and guidance in the role of reference materials in method development and validation

Received: 2 November 2002
Accepted: 3 February 2003

Presented at the CERMM-3, Central European Reference Materials and Measurements Conference: The function of reference materials in the measurement process, May 30–June 1, 2002, Rogaška Slatina, Slovenia

N. Majcen
Republic of Slovenia-Ministry of Education, Science and Sport, Metrology Institute (MIRS), Celje, Ljubljana, Slovenia

N. Majcen (✉)
Metrology Institute of the Republic of Slovenia (MIRS), Laboratory Centre Celje, Tkalska 15, SI-3000 Celje, Slovenia
e-mail: nineta.majcen@gov.si
Tel.: +386-3-4280754
Fax: +386-3-4280760

Abstract The fact that various definitions and terminology applied to measurements in analytical chemistry are not always consistent and straightforward, by not only answering the question "what", but also "how", leads to their various interpretations. This results in non-uniform implementation of very basic and essential metrological principles in chemistry. Such a diverse situation is not conducive to the endorsement of harmonised measurements all across the world, to serve as a tool for improving the quality of life in its broadest sense for all its citizens. The discussion in this paper is focused on problems associated with terminology and definitions of 'reference material' and 'validation'. The role of reference materials in measurement processes for purposes other than calibration and validation principles in analytical chemistry are also discussed in this paper. Where possible, potential solutions are proposed, but more often, questions of essential importance are raised in order to initiate international discussion which will hopefully lead to equally understandable answers.

Keywords Reference materials · Analytical chemistry · Measurements in chemistry · Validation · Method development

Introduction

Internationally understandable terminology and interpretation of definitions play an essential role in the analytical community worldwide [1–3]. Language, as a tool for communication, can considerably accelerate the unification of activities or alteratively act as a diversifying element that could keep analysts going on in various directions and not approaching the same goal. As this effect is intensified due to the various 'dimensions' of different languages that we speak, it is even more important to make the original definition as non-confusing as possible and to provide additional explanations in written and oral form to make sure that it is implemented properly. It is thus important, that the terms we use are 'traceable' to the same definition and its understanding/implementation in practice as this will result in comparable measurement results.

Only on the basis of internationally accepted definitions and terminology, can the intended uses of reference materials be discussed in a way uniformly understandable across all sectors where analytical chemistry is applied. While the production of reference materials and their storage is very well described in the literature, their intended use is somehow overlooked [4–13]. The consequence is that the analysts are left to their own imagination for how to use them, which as a boomerang (again) leads to different interpretations of results measured in various laboratories.

In order to make the exchange of information, based on results of measurements obtained in analytical laboratories reasonable, measurements should be traceable to the same stated unit and thus be comparable. Neither establishing nor demonstrating the traceability of measurement results is an easy task. Moreover, there is no general rule that can be proposed, which is valid and implemented in general. Validation is thus an important way to investigate and describe the traceability and uncertainty issue of the concrete measurement result. Since, up to now it has mainly not been considered as such, some updates related to interpretations of valid definitions are further discussed in the paper.

The problems associated with terminology

Reference material

As written in the Eurachem Guide 'it is commonplace to confuse reference material with certified reference material' [14]. Unfortunately, the situation is even worse, as there are also some other terms used to describe the same, like 'standard material', 'reference standard', 'standard reference material' etc. It is evident from the literature that in the International Vocabulary of General Terms in Metrology (VIM), at least twelve different terms associated with objects used to define, realise, conserve or reproduce a unit or a value of a quantity are used [1, 15]. Up to now several approaches have been already made to uniform terminology in this field as it is essential for many reasons (traceability and consequently comparability being one of them) and it is hoping that

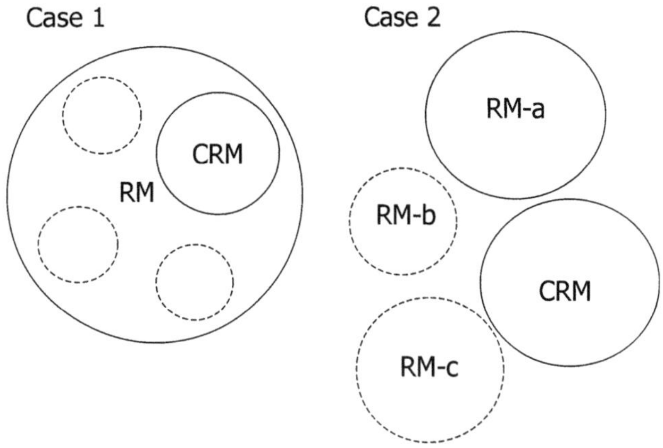

Fig. 1 Terminology: reference material (RM) vs. certified reference material (CRM). Inconsistent terminology is a fertile base for various interpretations of definitions: definitions given in Table 1 can be interpreted as given in case 1 (CRMs being a part of a 'family' named RM) or as given in case 2 (several different types of RM, one of them being CRM). Intended use of reference standard depends on accepted interpretation of definitions

the new revision of VIM which is due soon will make the situation better. Some of the most often used definitions are given in Table 1.

As reference materials can serve different purposes, it is of essential importance to agree on the terminology and definitions for each specific term. As it stands now, the definitions might be interpreted in at least two different ways, as shown on Fig. 1. If the definitions are un-

Table 1 Reference material: definitions

Document	Term	Definition
VIM [1]; ISO Guide 30:1992 [4]	Reference material	Material or substance one or more of whose property values are sufficiently homogeneous and well established to be used for (a) the calibration of an apparatus, (b) the assessment of a measurement method, or (c) for assigning values to materials
VIM [1]; ISO Guide 30:1992 [4]	Certified reference material	Reference material, *accompanied by a certificate*, one or more of whose property values are certified by a procedure which establishes traceability to an accurate realization of the unit in which the property values are expressed, and for which each certified value is accompanied by an *uncertainty* at a stated level of confidence. NOTE (4 of 5): All CRMs lie within the definition of *"measurement standards"* or *"etalons"* given in the VIM.
VIM [1]	Measurement standard, etalon	Material measure, measuring instrument, *reference material* or measuring system intended to define, realize, conserve or reproduce a unit or one or more values of a quantity to serve as a reference. EXAMPLES (2 of 6): standard hydrogen electrode; reference solution of cortisol in human serum having a certified concentration.
EURACHEM Guide on Validation, 1998 [14]	Reference material	RM can be virtually any material used as a basis for reference, and could include laboratory reagents of known purity, industrial chemicals, or other artifacts. The property or analyte of interest needs to be stable and homogenous but the material does not need to have the high degree of characterisation, traceability and certification more properly associated with CRMs.

Table 2 Validation: definitions

Document	Definition
EURACHEM Guide on Validation, 1998 [14]	Method validation as being the process of defining an analytical requirement, and confirming that the method under consideration has performance capabilities consistent with what the application requires.
ISO 9000:2000 [16]	Confirmation, through the provision of *objective evidence* (3.8.1.), that the *requirements* (3.1.2) *for a specific intended use or application* have been fulfilled NOTE 1 The term "validated" is used to designate the corresponding status NOTE 2 The use conditions for validation can be real or simulated.
ISO 17025:1999 [17]	Validation is the confirmation by examination and provision of objective evidence that the particular requirements for a specific intended use are fulfilled.

derstood as given in case 1, then reference materials can be used for calibration as well. If this is not the case, which is implied in case 2, then reference materials should not be used for calibration purposes. The problem is, of course, much more complicated, but is beyond the scope of this paper.

Method development

From literature and in practice, any change in analyte, matrix or both is described as 'method development' (better: 'procedure development'). Furthermore, this term also describes minor or major changes of part of the measurement (analytical) procedure e.g. a different dissolution step. Nowadays, it rarely happens that a certain measurement procedure is developed from scratch, as a completely new, i.e. where a certain 'new' technique is applied to measure a 'new' analyte in a 'new' matrix. However, there is no official definition in any of international documents of the term 'method development' and it seems the situation may remain as it is, since it does not cause confusion, misunderstandings and/or misinterpretation.

Procedure validation

The situation is becoming more complicated and less uniform when discussing validation. Several definitions (Table 2) are offered to analysts [14, 16–17]. The EURACHEM guide also gives some additional explanation of how the definition could be understood as well as guidance on how it might be implemented in practice. It is important to emphasise that other interpretations are also possible and that by no means the ones given in the above stated Guide are the best by definition. Additionally, it should be stressed that term 'procedure validation' instead of 'method validation' is used in this paper as this is more exact wording for how it should be understood within this context and what is actually taking place in practice.

Table 3 Validation: list of possible performance parameters to be evaluated and properties of the results obtained via validation

Performance parameters	Property
Working range, linearity	Traceability
Sensitivity, detection limit	Uncertainty
Repeatability	
Reproducibility	
Recovery	
Robustness	

If in the past, the procedure validation tended to concentrate on the process of evaluating the procedure's performance capabilities, the modern approach also confirms that the procedure under consideration has performance capabilities consistent with what the application requires. Implicit in this is that it will be necessary to evaluate the procedure's performance capabilities.

It therefore depends on each specific case, which performance parameters are to be confirmed before starting the analysis. The scope of validation thus might depend to a certain extent on the customers' needs. If they are very diverse (e.g. the laboratory has a customer, who is interested in lower concentrations of a particular analyte as well as a customer who brings a sample with a higher concentration of the same analyte, but this customer requires very low measurement uncertainty), it is worth broadening the scope of validation or confirmation, otherwise it would be sufficient to evaluate/confirm key parameters for the intended use. The general rule should be that only those parameters that are needed to be fulfilled due to the customers' requirements must be confirmed. However, a lot of additional measurements are usually done to evaluate the procedure's performance parameters (Table 3) and the result's properties (traceability, measurement uncertainty).

Intended uses of reference materials for quality purposes, including validation

Due to the above mentioned problems related to terminology it is essential to decide on the nomenclature that will be used within this paper in order to be in a position to give a clear explanation on purposes for which reference materials are used. Thus 'reference materials' are used as a 'family name' i.e. as given in case 1 on Fig. 1. As such, reference materials can be used for (a) establishing traceability and (b) quality assurance, which is the issue of this paper. The following aspects of quality assurance are covered by reference materials: (a) validation, (b) internal quality control, (c) external quality assessment [18–20].

Validation as one of the milestones of ISO/IEC 17025 is required to be performed each time non-standard methods, procedures developed in-house or methods outside their intended scope are used. Additionally, according to the same standard, every laboratory should confirm that it can properly operate standard or validated methods (procedures) before (first) introducing the tests, which, as a consequence again include the validation principle. The scope of validation ('objective evidence') i.e. which procedure's performance parameters are to be evaluated depends on the specific intended use e.g. (a) compliance with regulations, (b) maintaining quality and process control, (c) making regulatory decisions, (d) supporting national and international trade, (e) supporting research. As recommended by ISO/IEC 17025, validation and/or confirmation of the procedure should be done through uncertainty evaluation that is made through systematic assessment of the quantities influencing the result. This can be done either via measurement of the reference materials or by other means e.g. participation in interlaboratory comparison or comparison of results achieved with other procedures [21].

Various types of control charts, a tool that is most often used for internal quality control purposes, are often based on values given by reference materials. For this purpose, in-house reference materials should be preferably used, as certified reference materials are too expensive to be used to monitor the stability of the measurement process.

For external quality assessment purposes, a reference material with an undisclosed value is distributed to the participants to demonstrate their technical performance and skills. Comparison of the results of a certain laboratory with a reference range is a 'measure' of how good the laboratory is performing in a specific field. However, participation in interlaboratory comparisons (ILC) should be a regular activity and the results of regular participations should be traced and compared in time. It is important to be aware that demonstrating the improvement of performance is the key idea of external quality assessment, thus there is no need of being afraid of having 'bad' results in an ILCs.

Conclusion

Terminology and definitions related to reference materials and validation as well as their implementation in practice has been discussed in this paper. Several attempts are going on at the international scene to harmonise the definitions and their implementation in practice, which should, hopefully lead to comparable measurements in chemistry on an international scale. However, as the topic is very broad in its scope and extremely complicated, progress seems to be slower than desired, but would expectedly lead to complementary and consistent definitions. Therefore, it is not yet possible to give definite answers to all the questions stated in this paper as they are still under discussion at international level, but nevertheless, the following conclusions can be given.

1. There is an urgent need to harmonise terminology related to reference materials. Inconsistency in this field has a tremendous effect on all measurements in analytical laboratories as written standards, analytical measurement procedures and other documents are understood in different ways. Consequently, this often leads to results, which are not comparable as they are not traceable to the same stated unit.
2. The role of reference materials for quality purposes seems to be well implemented, but people are often not aware of their possibly different role in the calibration process, which might have important consequences with regard to the result of measurement. Thus, training of practitioners and education of students in the topics described in this paper are highly desirable.
3. Analysts and end users of the measurement results should be aware of the new dimension of the procedure validation definition given in ISO/IEC 17025, which requires that a procedure's performance parameters are fit for a specific intended use. In other words this means that the work of an analyst is not finished when performance capabilities of a specific method (or preferably 'procedure') are evaluated, but he/she has to go a step further and check whether these characteristics are in compliance with the client's needs. Of course, it is up to the client to specify his/her requirements/properties the result should have. Furthermore, ISO/IEC 17025 is introducing evaluation of measurement uncertainty as a mean of performing validation through systematic assessment of all quantities influencing the result.

Acknowledgements The author is grateful to Aleš Fajgelj for his comprehensive comments on the topic described in this paper. Sincere thanks also to Philip Taylor, Ewa Bulska, Emilia Vassileva, Miloslav Suchanek and Margreet Lauwaars for their contribution during fruitful discussions on validation.

References

1. International Vocabulary of Basic and General Terms in Metrology, BIPM, IEC, IFCC, ISO, IUPAC, IUPAP, OIML (1993)
2. Holcombe D. (2000) Accred Qual Assur 5:159–164
3. Terms and definitions used in connection with reference materials, ISO Guide 30 (1992)
4. Calibration in analytical chemistry and use of certified reference materials, ISO Guide 32 (1997)
5. Use of certified reference materials, ISO Guide 33 (2000)
6. The role of reference materials, Achieving quality in analytical chemistry, ISO Information booklet (2000)
7. The use of CRMs in the fields covered by metrological control exercised by the national service of legal metrology, OIML D18 (2002)
8. Stoeppler M., Wolf W.R., Jenks P.J. (Eds) (2001) Reference materials for chemical analysis, certification, availability and proper usage, Wiley-VCH
9. Fajgelj A., Parkany M. (Eds) (1999) The use of matrix reference materials in environmental analytical processes, RSC
10. De Bièvre P. (2000) Accred Qual Assur 5:307
11. De Bièvre P. (2000) Accred Qual Assur 5:224–230
12. Fajgelj A. (2000) Spectroscopy 15(6): 19–21
13. Günzler H. (Ed) Accreditation and quality assurance in analytical chemistry, Springer
14. The fitness for purpose of analytical methods, A laboratory guide to method validation and related topics, EURACHEM (1999)
15. Örnemark U, Discussion paper on terminology associated with reference material, 4E/RM (2001)
16. Quality management systems – Fundamentals and vocabulary ISO 9000 (2000)
17. General requirements for the competence of testing and calibration laboratories, ISO/IEC 17025 (1999)
18. Neidhart N., Wegscheider W. (Eds) (2001) Quality in chemical measurements, Springer
19. Harmonised guidelines for the use of recovery information in analytical measurement, IUPAC, ISO, AOAC, EURACHEM, Technical Report (1998)
20. Marschal A., Andrieux, Compagon P.A., Fabre H. (2002) Accred Qual Assur 7:42–49
21. TrainMiC material (2002) www.trainmic.org

Accred Qual Assur (2003) 8:235–241
DOI 10.1007/s00769-003-0624-x

Eugenija Ramoškienė
Mykolas Gladkovas
Mudis Šalkauskas

Validation of salt spray corrosion test

Received: 14 September 2002
Accepted: 11 March 2003
Published online: 23 April 2003

Presented at the Conference "Metrology and Measurement – 2001", 24–25 April 2001, Kaunas, Lithuania

E. Ramoškienė (✉) · M. Gladkovas
M. Šalkauskas
Institute of Chemistry,
Goštauto, 92600 Vilnius, Lithuania
e-mail: chemins@ktl.mii.lt; mudis@mail.lt

Abstract Quality control of corrosion test results implies the validation of the corrosion test method and estimation of the uncertainty of corrosion rate measurement. The corrosion test in an artificial atmosphere of the salt spray mist needs evaluation of corrosivity of the test cabinet by reference specimens. Such calibration of corrosion environment raises very strict requirements for the method description and details of all procedures and used specimens. Reliable corrosion measurements by spray tests require validation of the experimental device together with the experimental procedure and determination of corrosivity uncertainty of the test cabinet environment.

Corrosion tests have been conducted for a long time but there are only a few cases of corrosion data quality assessment or interlaboratory comparisons for such measurements. Each test method when used in different laboratories gives different results, as it is impossible to perform the whole procedure exactly in the same manner. Therefore, a very essential parameter of the method is its robustness. A proper validation of the corrosion test method means the evaluation of the impact of various environmental features and performance variations on the uncertainty of the test result.

Our aim was to present an experimental evaluation of the corrosivity of the salt spray corrosion test cabinet, to indicate the gaps in the description of the corrosion test method according to ISO 9227 and to estimate the main components of the uncertainty of the corrosivity measurement.

The validation results require changes in the salt spray test method description and maybe in the performance.

Keywords Validation of method · Corrosivity · Measurement uncertainty

Introduction

Corrosion tests in artificial atmospheres [1, 2, 3] are used as comparative tests for the evaluation of corrosivity of metals and metal alloys and corrosion protection capability of various corrosion protection means by metal plating, varnishing and paint coating as well as anodic and conversion coating. Therefore, it is essential to know precisely the corrosivity of the test cabinet environment.

The standard method for the test cabinet corrosivity determination is described in ISO 9227[1] [1] but we failed to find any information about the validation of this method or its metrological parameter evaluation. On the other hand, it is necessary to determine from the experimental point of view whether this technique is reliable enough as a standard method. A procedure is described in ISO

[1] Translated and accepted as LST – national standard of Lithuania Republic.

5725-1 [4] for the estimation if the standard method is sufficiently detailed and can possibly be improved.

ISO 9227 [1] does not specify in detail many necessary parameters and does not determine the precision of such a test method. The precision and accuracy of corrosion determination are influenced by many factors: preparation of specimens, conditioning, removal of corrosion products, cleaning, drying, etc. In literature on the corrosion tests we failed to find any information concerning the quality of corrosion tests results. The aim of this paper is to call attention to the problems in the corrosion measurement data quality and the necessity to evaluate the uncertainty for measurement results. We attempted to show the main components of uncertainty of the result in such a measurement on the basis of the experimental evaluation of the corrosivity of the spray test corrosion cabinet by means of reference specimens.

Materials and methods

An accuracy experiment can often be considered as a practical test of the adequacy of the standard measurement method. One of the main purposes of standardization is to eliminate differences between users (laboratories) as far as possible and the data provided by the experiment should reveal how effectively this purpose can be achieved.

According to the ISO 5725-1 requirements [4] a description of the measurement method is one of the main sources of the uncertainty and therefore it is essential for traceability of results. Therefore, an analysis of the description of the standard test method was performed and some ambiguous statements or lack of information for experimental procedures were pinpointed as shortcomings. The details of the experiment, which do not meet the requirements of the description of the reference method, were marked off as well. The result of these attempts is presented in Table 1.

Evaluation of cabinet corrosivity. In order to determine the corrosivity of the corrosion cabinet environment eight tests were performed [5] according to the standard method of the neutral salt spray test (Table 1). The results of corrosion rate of RS and the main statistical parameters such as the number of reference samples n, average RS mass m and RS mass loss Δm of each RS, average RS surface area S and surface area of each RS S_n, mean averages of all eight experiments and their standard deviations are presented in Table 2a and 2b.

Corrosion rate values were calculated from the mass loss of each test presenting a statistical array and indicating corrosion rate ordered array average value v, mode value v_{mode}, median value v_{median}, and standard deviation s.

The statistical analysis of the data was performed for the determination of outliers by means of their interquartile range:

Table 1 Comparative analysis of standard LST ISO 92227:1997 and supplementary standards EN ISO 7384:1998 (1), LST ISO 7253:1998 (2) and possibilities of their experimental realization

Requirements of standard/reference methods	Peculiarities of the experimental device and procedures and shortcomings in the standard
Corrosion cabinet	
Cabinet capacity no less than 0.2 m^3 and preferably not less than 0.4 m^3	Cabinet capacity – 0.4 m^3
At least two fog collecting devices in the zones of the test specimens – glass cylinder with funnels with a diameter of 100 mm	Four collecting devices in the corners
Inert plastic supports for the test specimens	PMMA plastic support for the five specimens
Automatic registration of temperature and air humidity (1)	Not provided automatic registration[a]
Determined humidity should be kept within ±5% (1)	No data about humidity value
If necessary, air circulation system shall be provided	Not provided[a]
Spraying device	
Supply of clean air of controlled pressure [(70 to 170)±0.7] kPa	Compressed air pressure within determined interval, and supplied without filtration
Air shall be humidified in saturation tower at temperature several degrees higher than that of the cabinet	Not provided[a]
Level of the water must be maintained automatically	Provided
Atomizers shall be made of inert material	PMMA plastic
Adjustable baffles to obtain uniform distribution of the spray	PMMA plastic
Level of the salt solution shall be maintained automatically	Maintained automatically
Test solution	
Sodium chloride (50±5) g in distilled or deionised water with a conductivity not higher than 20 µS/cm at (25±2) °C	The test solution prepared from analytical grade NaCl (GOST 4233-77); conductivity of deionised water not checked
NaCl shall contain less than 0.001% of Ni, 0.001% Cu, 0.1% of NaI and 0.5% (m/m) of total impurities; specific gravity range of the solution is 1.0255 to 1.0400 and the pH range within 6.0 to 7.0 at (25±2) °C	Cu and Ni determined by AAS and both met the requirements of ISO 9227; the pH of the prepared solution measured with millivoltmeter pH-121

Table 1 (continued)

Requirements of standard/reference methods	Peculiarities of the experimental device and procedures and shortcomings in the standard
Reference specimens (RS)	
RS cut from cold-rolled plates or strips of CR4 grade steel (according to ISO 3574) (1±0.2) mm thick and (50×80) mm with faultless surface and a mat finish R_a=1.3±0.4 µm	RS of stated parameters cut from DC 04 AMO (EN 10130) steel plates (800×1000) mm; not clear what "mat finish" means; our surface R_a=0.67±0.4 µm
Cut edge should not be sharp	RS marked on the back side by stamping with numbers and their cut edges grounded
	Length and widths of RS measured with a vernier (graduation value 0.05 mm)
	Thickness measured with micrometer MK-0-25 mm (GOST 6507-78) (accuracy 0.01 mm) Storage of RS before pretreatment not specified
Pretreatment of RS before test	
Clean RS immediately before use in a vessel full of an appropriate organic solvent (hydrocarbon, with a boiling point between 60 °C and 120 °C) using a clean soft brush or an ultrasonic cleaning device	RS cleaned before use (not clear what "immediately" means) in hexane, b.p. 64 °C
Rinse RS with fresh solvent, then dry them	Rinsed with hexane and dried by fan
Keep RS in a desiccator with a proper drying agent for 24 h (1) or dry RS over proper time interval at proper conditions (2)	RS kept in a desiccator 30 min to 1 h; not clear what "proper drying agent" and "proper time" means
Weigh RS a to ±1 mg	RS weighed using second precision class laboratory balances VLP-200 g
Protect one face of RS with removable coating	Adhesive plastic film used
Arrangement of the RS	
Position each one RS on four different quadrants with the unprotected face upwards at an angle of 20°±5° from the vertical	Five RS on every cabinet quadrant according to the standard requirements
The upper edge of RS shall be level with the top of the salt spray collector	Salt spray collectors ca. 5 cm lower than the upper edge of RS
Operation condition in the spray cabinet	
Temperature should be (35±2) °C and shall be measured at least 100 mm from the walls	Requirement met
The average rate of collection of solution in each collecting device, measured at least over 24 h of continuous spraying, shall contain 1 ml/h to 2 ml/h and shall have a sodium chloride concentration of (50±5) g/l and pH value in the range of 6.5 to 7.2	Requirement met
The test duration is 96 h	Requirement met
Treatment of RS after the test	
Remove the protective coating	The mode of removal not specified; the protective coating stripped off
Remove the corrosion products by immersion in cleaning solution at 20 °C to 25 °C for 10 min (according to ISO 8407 [8])	RS rinsed with cool running water with a soft sponge, then it dipped for 10 min into the cleaning solution (500 ml hydrochloric acid and 3.5 g hexamethylenetetramine; distilled water to make 1000 ml) prepared according to ISO 8407
Thoroughly clean the RS at ambient temperature with water, then with acetone	Requirement met
Followed by drying	Duration not specified; RS dried using fan for (5–10) min and kept in a desiccator
Weigh the RS to ±1 mg and calculate the mass loss in g/m^2	After ca. 30 min weighed according to the requirements
Evaluation of cabinet corrosivity	
Operation of the test apparatus is satisfactory if the mass loss of each reference specimen is (140±40) g/m^2	Ambiguous meaning of "±40 g/m^2" may be means confidence interval and there is no statement on uncertainty of result value

[a] As far as it was beyond the ability of our equipment and laboratory provision

Table 2a Primary data (n) of salt spry test cabinet corrosivity evaluation in eight experiments as corrosion rate (v, mg/m^2) of RS (average mass m=31225 mg) mass loss (Δm, mg) from their geometric surface area (S, mm^2) (first four experiments)

No of experiment	1			2			3			4		
Raw data	Δm	S	v	Δm	S	v	Δm	S	v	Δm	S	v
1	0.348	3972	87.7	0.408	4054	100.6	0.391	3957	98.0	0.382	3982	96.0
2	0.390	3998	97.6	0.426	4102	103.9	0.393	4004	98.0	0.434	4071	106.6
3	0.403	4000	100.8	0.425	4090	103.9	0.413	4032	102.5	0.454	4096	107.6
4	0.403	3980	101.3	0.423	4054	104.2	0.413	3988	103.2	0.441	4038	109.3
5	0.403	3976	101.5	0.431	4129	104.3	0.425	3990	106.5	0.452	4073	110.0
6	0.406	3980	102.0	0.430	4102	104.7	0.430	4038	106.6	0.445	4026	110.5
7	0.404	3953	102.2	0.432	4085	105.7	0.429	4014	106.9	0.457	4096	111.7
8	0.410	3969	103.4	0.432	4074	106.0	0.434	3985	108.9	0.468	4176	112.1
9	0.421	3982	103.8	0.435	4060	107.2	0.436	3988	109.2	0.457	4073	112.2
10	0.417	4009	104.1	0.439	4070	107.8	0.437	3991	109.5	0.439	4049	112.4
11	0.416	3991	104.3	0.440	4066	108.2	0.434	3955	109.9	0.471	4186	112.6
12	0.419	3988	105.0	0.441	4053	108.8	0.446	4023	110.8	0.452	4002	112.8
13	0.421	3997	105.4	0.453	4116	110.1	0.446	4018	111.0	0.469	4091	114.6
14	0.428	4019	106.4	0.440	3993	110.3	0.447	4026	111.1	0.474	4120	114.9
15	0.427	3988	107.1	0.455	4109	110.7	0.443	3958	112.1	0.470	4083	115.1
16	0.429	3997	107.4	0.452	4079	110.9	0.478	4138	115.4	0.469	4056	115.6
17	0.436	3996	109.2	0.436	4056	110.9	0.457	3943	116.0			
18	0.433	3947	109.7	0.460	4082	112.7	0.510	3960	128.8			
19	0.444	4029	110.2	0.466	4083	114.1						
20	0.477	3992	119.5	0.465	4070	114.3						
Average	0.417	3988.2	104.4	0.439	4076.4	108.0	0.437	4000.4	109.1	0.452	4076.1	110.9
Standard deviation	0.02	19.2	5.9	0.01	28.7	3.7	0.03	43.7	6.8	0.02	52.8	4.6

IQR=Q_3–Q_1, where Q_3 is the third quartile and Q_1 is the first quartile, when the median divides the experimental sample into two parts. Outliers were found in experiments No 1, 3 and 6 and may be rejected but we did not find enough good reason for that and they were used in calculations of the average corrosion rate value and its standard deviation.

There is rather significant scattering of data and some of the eight experiments may be regarded as outliers. Therefore we have tested the null hypothesis H_0 of equality of the lowest mean of corrosion rate in experiment No 1 and the highest mean in experiment No 8. The calculated value $F=s_1^2/s_8^2=5.9^2/5.0^2=1.4$ for standard deviations was compared with Fisher distribution statistical test values

$$F_{1-\alpha/2}(n_1-1,n_8-1)\le F\le F_{1-\alpha/2}(n_1-1,n_8-1),$$

where α is significance level 0.1, and n_1 and n_8 number of RS in tests No 1 and No 8.

Null hypothesis is acceptable as $F_{1-0.05}(19,19)=0.5\le F=1.4\le 2.02=F_{0.05}(19,19)$. The null hypothesis of equality of means of equal standard deviations is not acceptable when

$$t=(v_8-v_1)\sqrt{v(n_1n_8(n_1+n_8-2)/(n_1+n_8)}/$$

$$\sqrt{(n_1-1)s_1^2+(n_8-1)s_8^2}=3.29>t_{a=0.05}(38)=2.7.$$

Ordinary statistical analysis of the corrosivity tests showed that the mean of corrosion rate v=111 g/m^2 of eight experiments means and its standard deviation s=4.1 g/m^2 estimate of neutral salt spray test cabinet corrosivity as 111±3 g/m^2 (confidence 95%) shows that corrosivity of our corrosion cabinet meets the requirements of ISO 9227, which states that the average value and its data scattering should be 140±40 g/m^2.

For the total number of tested RS n=148, the mean of corrosion rate v=112 and its standard deviation s=7.5 differs insignificantly from the mean of the result of eight experiments which must be regarded as separate tasks each of which may have specific influence on the result. Null hypothesis of equality of means when standard deviations is unequal may be rejected when $t>t_{\alpha/2}(k)$ and in our case $t=0.64<t_{0.05}(45)=1.684$.

For more advanced analysis it is necessary to calculate the influence of many other factors and express it as the uncertainty of the result. The next part of this article deals with this task on the basis of our experimental data.

Estimation of corrosivity uncertainty. The corrosion rate (v) of reference specimens (RS) at the stated conditions according to ISO 9227 in the neutral salt test cabinet may be expressed as follows:

$$v=\Delta m/S\cdot t,$$

where Δm is mass loss after the test/ during the test, S is area of reference specimen and t *is* duration of the corrosion test (according to the standard method the test duration is 96 h).

Therefore, the corrosivity of the cabinet was evaluated as $v=\Delta m/S$ g/m^2 for 96 h in the eight experiments. The analysis of the main uncertainty sources according to recommendations [6, 7] was performed (Fig. 1) for evaluation of possible value of uncertainty of corrosivity measurement result. Each main source of uncertainty (mass loss Δm, surface area S and duration t) was analysed and calculated separately and these components used for combined and expanded uncertainty calculation.

Uncertainty of mass loss measurement. The standard method of the neutral salt spray test does not indicate the mass of RS. Mass loss was found as a difference between the RS prepared for the corrosion test and the RS after the corrosion test and corrosion product stripping as well as protective coating removal from the RS (Table 1). Such a mass loss determination is based on three components: (1) mass loss determination by weighing (accuracy 0.5 mg and standard deviation 0.3 mg) before the neutral salt spray test and after it, (2) determination of difference and (3) cor-

Table 2b Primary data (n) of salt spry test cabinet corrosivity evaluation in eight experiments as corrosion rate (v, mg/m^2) of RS (average mass m=31225 mg) mass loss (Δm, mg) from their geometric surface area (S, mm^2) (continuation—second four experiments)

5			6			7			8		
Δm	S	v	Δm	S	v	Δm	S	v	Δm	S	v
0.415	4022	103.2	0.404	4092	98.7	0.425	4017	105.7	0.455	4067	111.9
0.425	4005	105.5	0.421	4094	102.8	0.430	4160	103.4	0.464	4012	115.6
0.425	3968	106.7	0.427	4105	104.1	0.415	3968	104.7	0.496	4000	117.2
0.424	3972	106.8	0.417	3960	105.4	0.423	3994	105.9	0.468	3997	117.3
0.437	4042	108.2	0.437	4097	106.7	0.437	3986	109.6	0.468	3981	117.5
0.435	3982	109.2	0.446	4096	108.8	0.439	3969	110.7	0.467	3967	117.6
0.444	4064	109.2	0.450	4107	109.6	0.443	3950	112.2	0.479	3977	119.9
0.436	3966	110.0	0.449	4081	109.9	0.454	3995	113.6	0.485	4014	120.8
0.442	3963	111.6	0.453	4099	110.4	0.459	3985	115.3	0.483	3985	121.2
0.448	3966	112.1	0.452	4082	110.8	0.460	3976	115.6	0.491	3999	122.8
0.458	4027	113.8	0.453	4001	113.2	0.464	4009	115.7	0.502	4059	123.8
0.460	4018	114.4	0.472	4125	114.4	0.463	4003	115.8	0.499	4014	124.4
0.455	3971	114.5	0.479	4084	117.3	0.466	3986	116.8	0.496	3985	124.5
0.461	4011	114.9	0.487	4099	118.9	0.471	3957	119.0	0.500	4005	124.8
0.458	3953	116.1	0.484	4046	119.7	0.492	3945	124.8	0.498	3964	125.6
0.462	3964	116.6	0.499	4130	120.9	0.501	3993	125.6	0.501	3971	126.1
0.470	4018	117.1	0.492	4056	121.2				0.522	4061	128.5
0.467	3960	117.9	0.500	4098	122.1				0.521	4034	129.1
0.481	4017	119.7	0.513	4069	126.1				0.515	3973	129.7
									0.517	3971	129.8
0.448	3994.2	112.0	0.460	4080.1	112.7	0.453	3993.3	113.4	0.491	4001.8	122.4
0.02	31.8	4.5	0.03	40.0	7.3	0.02	47.4	6.4	0.02	31.2	5.0

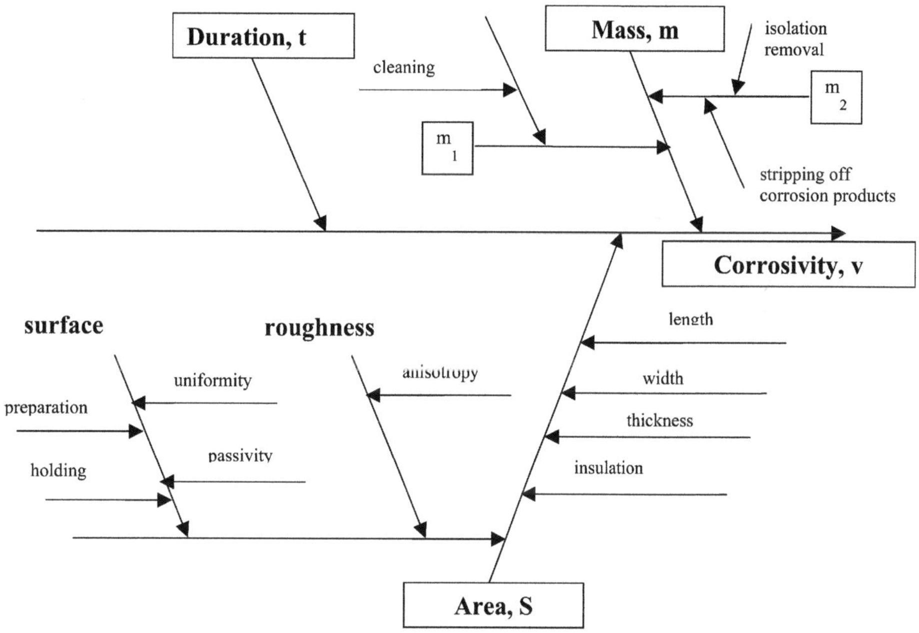

Fig. 1 Main uncertainty sources of corrosivity measurement as rate of mass loss in a neutral salt spray cabinet expressed as $v=\Delta m/S$

rosion product stripping (accuracy 0.5 mg and standard deviation 0.9 mg).

Five samples were cleaned, dried and weighed (with ±1 mg accuracy) for the determination of RS unprotected surface dissolution in the corrosion product stripping solution, containing the hydrochloric acid. After 10 min dripping in the hydrochloric acid solution they were rinsed at room temperature with water, acetone and dried before weighing. The average mass loss was 2 mg and its standard deviation was 0.9 mg.

The value of such a systematic error, if used for the result correction, diminishes the mass loss value and influences the standard uncertainty of the result with its standard deviation (in our case 0.9 mg). Generally, it is possible to use relative uncertainty u_c=0.4.

The uncertainty of the difference of the dissolution rates in the corrosion product stripping solution between the corroded surface and the surface protected from corrosion but unprotected in the stripping solution corresponding surface was not determined. A special investigation should be conducted and instructions added to the method documentation for proper uncertainty determination. The calculated standard uncertainty $u(\Delta m)$=0.027 g of mass loss 0.4 g determined as an average from 148 reference specimens after corrosion in the neutral salt spray cabinet within 96 h.

Uncertainty of surface area determination. According to the description of the standard test method, the RS should be plates (50×80×1 mm); the total geometric surface area of such samples is 8260 mm^2. The corrosion test in neutral salt spray environment is performed only on one side of the plate and the other side of the plate is protected by an adhesive plastic film; thus the treated geometric surface area is reduced to 4260 mm^2. About 6% of this area is the surface of plate edges, the structure of which depends on the plate formation technique (cutting, slashing, clipping etc.). Thus, the corrosion rate of such a surface may be different from the plane surface.

Insulation tightness and adherence of the adhesive plastic film to the protected surface of the plate contributes to the total surface area under the corrosion test. It may be assumed that such unevenness of the surface protection occurs within a one millimetre wide zone along the whole perimeter and it makes up ca. 6% of the total geometric surface area.

The measurement accuracy of the geometric surface area should be evaluated as ±2.2 mm^2 according to the standard method where it is stated that RS must be 50×80 mm and it may be interpreted as a measurement with accuracy ±0.1 mm.

For the uncertainty determination of the geometric surface area, the calculations of combined uncertainty were performed on the basis of the experimental measurements and the analysis of a full set of parameters (up to 10); each of them may contribute to the combined uncertainty of the total geometric area of the surface of the test plate according to the recommendation [7]. Experimentally we estimated only seven parameters contributing to uncertainty: geometric surface area, length, width, thickness, completeness/evenness of insulation, surface roughness and its anisotropy [5].

The geometric surface area is only a basic parameter for the corrosion rate expression in normalized unit g/m^2 within normalized test duration. An actual surface area is mentioned [2], but not defined in standards [1, 2]; therefore we use the geometric surface and try to evaluate influence of various factors that increase the real surface area. The corrosion rate depends on such surface parameters as surface roughness and its anisotropy, evenness of material composition on the surface, pretreatment of surface, storage of samples before the corrosion test, and so on. Inasmuch as these parameters have no proper units for quantitative estimation expression, it was impossible to evaluate their contribution to the combined uncertainty of the tested surface area. The nominal value of surface roughness in the standard method is indicated as R_a=1.3±0.4 µm, but in our case R_a=0.67±0.01 mm, with an anisotropy R_{ai}=0.60 µm (along the plate) and R_{as}=0.74 µm (across the plate).

Combined uncertainty of the geometric surface area, calculated from the experimental data value, is 2.4×10^3 mm^2 or 2.4×10^{-3} m^2. It is of the same order as that calculated from the data given in the standard method, but the expanded uncertainty is twice as large as is the combined uncertainty according to the indicated precision of measurements in the standard method. Thus, the multi-measurement total data scattering is larger than the uncertainty of measurements of a single plate.

The most significant components of the surface area uncertainty are the thickness of the corrosion test plate and the roughness of its surface; anisotropy may give the contribution up to 90%.

Uncertainty of corrosion duration. In the standard method normal corrosion test duration is indicated as 96 h, and this means that the measurement accuracy is ±1 h. Our experiments were performed with ±0.2 h accuracy since the corrosion cabinet was opened for several minutes every 24 h to check the quantity of the collected mist. These interruptions were not included in the corrosion duration. The duration of the corrosion test in the neutral salt spray cabinet was measured from the moment when the RS, placed in the cabinet, reached the indicated corrosion conditions.

Table 3 Uncertainty components (u—standard uncertainty, u_c—relative uncertainty) of corrosivity measurement evaluated as rate of mass loss in a neutral salt spray test cabinet

Component	Value[a]	u	u_c
Mass loss (change in mass), Δm	0.4 g	0.027 g	0.59
Geometric surface area, S	4×10^{-3} m^2	2.4×10^{-3} m^2	0.60
Duration of corrosion process, t	96 h	7.1×10^{-1} h	7.4×10^{-3}

[a] Data value are approximated to one significant digit

It was assumed that the uncertainty of corrosion test duration comprised two components: (1) accuracy which was indirectly indicated in standard method description and (2) our experimental possibility to keep the total duration of corrosion process in the neutral salt spray environment within limits of ±0.2 h.

The estimation of corrosion process duration (without any consideration of the peculiarities of the corrosion process in the test cabinet environment and experimental conditions) gives the value of 0.14 h.

Uncertainty of reference specimen corrosion rate. According to formula of specific uncertainty addition for corrosion rate calculation, its uncertainty has three components:

- Uncertainty of mass loss measurement
- Uncertainty of surface area measurement
- Uncertainty of corrosion process duration

Therefore, in our case (Table 3) combined uncertainty of corrosion rate $u_c(v)$ obtained from the corresponding relative uncertainties $u_c(\Delta m)$, $u_c(S)$ and $u_c(t)$ has the value 0.97:

$$u_c(v) = \sqrt{u_c(\Delta m)^2 + u_c(S)^2 + u_c(t)^2}$$
$$= \sqrt{0.59^2 + 0.6^2 + 7.4^2 \cdot 10^{-6}} = 0.97.$$

The expanded uncertainty U with a 95% confidence is U=111×2×0.97~215 g/m^2 for the experimentally determined average value of corrosion rate 111 g/m^2 within 96 h.

Results and discussion

Additional specifications from other corrosion test standards [2, 3] are required for the standard method of the accelerated corrosion test in the neutral salt spray test cabinet at 35±2 °C within 96 h. For the evaluation of corrosion data quality, the test should be performed according to the requirements of contemporary standards such as ISO/IEC 17025 [9] and, therefore, the corrosion test data uncertainty must be determined.

Experimental possibilities and technical conditions of the corrosion test cabinet allow us to present preliminary metrological parameters of such measurements. Many parameters are rather sensitive to some factors of corrosion environment conditions and performance procedure.

Experimental data reveal that the corrosivity of the test cabinet is greatly influenced by the procedure of performance, and by design parameters of the test cabinet. The quality of hydrochloric acid, used for the corrosion product stripping, as well as the other parameters of used materials, which should be checked by additional experiments, influence greatly the quality of experimental data too.

The analysis of experimental data shows that the average value 111 g/m^2 of all corrosivity data (improved by rejecting outliers) corresponds to the value 140±40 g/m^2 indicated in the standard. For the evaluation of the expanded combined uncertainty U with factor k=2 the corrosivity measurement gives the value of ±215 g/m^2 (at 95% confidence). It means that our data uncertainty is five-times higher than that specified in the standard as the data scattering interval ±40 g/m^2 and seven times as wide compared the statistic confidence interval in our own experimental data corrosivity (Table 2a and 2b). The main components of the combined uncertainty are mass loss and surface area determination.

The analysis of uncertainty components of the RS surface area shows that surface roughness (especially its anisotropy) greatly influences the resulting value. A disadvantage of the standard description is that there is no instruction how to manage surface roughness when its value differs from R_a=1.3±0.4 μm. The surface roughness of our specimens was ca. 0.7 μm, i.e. twice as small as stated in standard [1]; therefore, it is possible that average value of the established corrosivity 111±215 g/m^2 corresponds to the lower indicated limit of the standard value 140±40 g/m^2, i.e. 100 g/m^2.

Conclusions

Corrosion testing by salt spray standard procedure reveals that prescriptions given in ISO 9227 need some improvements in specifications and cross-references to provide a link with ISO 7384, ISO 7253.

The experimentally determined value of the spray cabinet corrosivity (in 96 h) 111±3 g/m^2 covers the value interval 140±40 g/m^2 indicated in the standard method. The standard deviation in experimental data scattering is rather small (s=4.1 g/m^2); the calculated expanded uncertainty (±215 g/m^2), however, exceeds interval indicated in the standard method.

The analysis of uncertainty sources and components reveals that the uncertainty value of the spray cabinet corrosivity depends on the measurements of the surface area of the RS and mass loss in the corrosion processes.

If it is necessary to minimize the uncertainty of the corrosivity result, the surface area determination and surface roughness specification should be improved as well as mass loss determination, which is based on a rather complicated procedure of corrosion product removal.

For a more comprehensive evaluation of corrosivity measurement uncertainty, it is necessary (1) to carry out an experimental investigation of the influence of each particular factor of corrosion test cabinet environment and experimental procedure and (2) construct a more detailed measurement model.

References

1. ISO (1990) Corrosion tests in artificial atmospheres – salt spray tests. ISO 9227, ISO, Geneva, Switzerland
2. ISO (1995) Corrosion test in artificial atmosphere – general requirements. EN ISO 7384, ISO, Geneva, Switzerland (ISO 7384:1986)
3. ISO (1996) Paints and varnishes – determination of resistance to neutral salt spray (fog). ISO 7253, ISO, Geneva, Switzerland
4. ISO (1994) Accuracy (trueness and precision) of measurement methods and results – Part 1: general principles and definitions. ISO 5725-1:1994/Cor.1:1998, ISO, Geneva, Switzerland
5. Ramoškienė E, Gladkovas M, Šalkauskas M (2001) Matavimai 2(18):13–17 (in Lithuanian)
6. EURACHEM/CITAC (2000) Quantifying uncertainty in analytical measurement, 2nd edn. EURACHEM/CITAC Guide, EURACHEM/CITAC, Teddington, UK
7. Barwick VJ, Ellison SL (2000) Protocol for uncertainty evaluation from validation data (LGC/VAM/1998/088). Teddington LGC, ; www.vam.org.uk
8. ISO (1991) Corrosion of metals and alloys – removal of corrosion products from corrosion test specimens. ISO 8407, ISO, Geneva, Switzerland
9. ISO/IEC (1999) General requirements for the competence of testing and calibration laboratories. ISO/IEC 17025, ISO, Geneva, Switzerland

Accred Qual Assur (2004) 9:253–258
DOI 10.1007/s00769-004-0767-4

Margreet Lauwaars
Elke Anklam

Method validation and reference materials

Received: 25 July 2003
Accepted: 27 November 2003
Published online: 5 February 2004

Presented at BERM-9—Ninth International Symposium on Biological and Environmental Reference Materials, 15–19 June 2003, Berlin, Germany.

M. Lauwaars · E. Anklam (✉)
European Commission,
Directorate General Joint Research Centre,
Institute for Reference Materials
and Measurements,
Food Safety and Quality Unit,
Retieseweg 111, 2440 Geel, Belgium
e-mail: elke.anklam@cec.eu.int
Tel.: +32-14-571316
Fax: +32-14-571783

Abstract For implementation of food and feed legislation, there is a strong need for development and harmonisation of reliable, validated and if possible, robust and simple analytical methods. In addition, precise methods used for measuring the exposure of humans to certain types of food contaminants and residues (natural, man-made or produced during technological treatment) such as, e.g. mycotoxins, acrylamide, pesticides and allergens have to be available, in order to compare results derived from monitoring studies. Methods should be validated (in-house or in a collaborative trial) according to harmonised protocols and good laboratory practice must be in place in order to be compliant with internationally harmonised standards. The way in which this is implemented depends strongly on the analyte, interference within the food matrix and other requirements that need to be met. Food and feed certified reference materials, when matrix matched and containing the appropriate concentration of the certified substance, are an extremely useful tool in validation of measurements.

Keywords Food analysis · Food safety and quality · Method validation · Certified reference materials

Introduction

Food safety and quality is an issue that concerns every citizen in the European Union and of course worldwide this is covered by the Joint FAO/WHO Codex Alimentarius Food Standards Programme. Food safety in its scientific meaning regards human health issues. Distinctions have to be made between food effects that impair the immediate health state of a person (i.e. cause illness within hours or days) and adverse effects which manifest themselves only after a prolonged period (months and years).

Food quality is one of the most important factors determining the consumer's perception and acceptance, attraction and purchase of a product. The consumer expects a wide range of competitively priced food products of consistently high quality. The market competition and consumer pressure to move away from the use of highly processed food products towards more "natural" food has motivated the food industry in the development of novel foods, ingredients and processes. Authenticity proof and detection of fraud, together with assessment of compliance with labelling, are therefore in the service of the consumer.

Demand-driven re-active and pro-active food control is crucial to ensure consumer protection. As animal feed is the prime source of contamination entering food its safety is also important to ensure the health of livestock and the safety of animal-derived food products.

Therefore, food and feed legislation has been put in place at both national and European levels for some time. For its implementation, there is a strong need for development and harmonisation of reliable, validated and if possible, simple analytical methods.

In the European Union, the authorities in the Member States that are competent to perform official controls shall meet operational criteria that guarantee their efficiency, effectiveness and impartiality [1]. The Directive on the subject of additional measures concerning the official control of foodstuffs [2] (to be amended by [1]) demands that official food control laboratories use validated methods of analysis, whenever possible. For this reason, analytical methods used by enforcement laboratories for the implementation of legislation must be subjected to validation procedures, in order to show that the method produces reliable results. Methods need to provide accurate, repeatable and reproducible results within and between different laboratories. This is extremely important in view of legal actions and trade specifications, as well as for monitoring or risk assessment studies. Method validation is done to check the performance of a method and assess its performance characteristics. The precision of the method is likewise important as a "fit for purpose" requirement.

Food and feed reference materials (FF-RMs) and especially certified reference materials (CRMs) play an important role in the verification of the accuracy of analytical measurements. They can be used as part of the measurement uncertainty estimation and to assess the traceability of the analytical result. CRMs are also used in several cases for the calibration of the instrumental set-up.

As food and feed commodities represent a very difficult matrix, it is a challenge for food analysts to be able to detect and quantify all kinds of food constituents, additives, residues and contaminants, including their metabolites, at all possible concentration levels. Moreover, sometimes the analytes are not even very well characterised, such as total fat, water, total carbohydrates and total proteins. These food constituents consist of a complex mixture of chemical substances (e.g. for fat it can be the sum of triglycerides, phospholipids, glycolipids and monodiglycerides and diglycerides). The type of method applied, in the case of so-called defining methods (e.g. drying oven method for water or determination of total nitrogen content for proteins), can define the substance(s) analysed. Another difficulty in food analysis is that the analyte may be strongly bound to the matrix, which influences the extraction efficiency. Transformation of original food constituents during food processing or storage may also occur. Some feed additives may form metabolites in animals and therefore those substances have to be looked for in animal-derived food products. There is a "cry" from the official food and feed control laboratories for many more CRMs, or at least test materials (TMs), than those existing to date.

Impact of validated methods and reference materials on implementation of food safety legislation

Official food laboratories

As laid down in European legislation [1], analysis of food samples taken during official controls shall be carried out by laboratories designated for that purpose by the competent authority. Any laboratory that is assessed and accredited in accordance with European Standards developed by the European Committee for Standardisation (CEN) shall be eligible for designation as an official laboratory. The standards are according to [3]:

- EN ISO/IEC 17025 on "General requirements for the competence of testing and calibration laboratories"
- EN 45002 on "General criteria for the assessment of testing laboratories"
- EN 45003 on "Calibration and testing laboratory accreditation system—general requirements for operation and recognition".

The ISO 17025 standard [3] describes monitoring the quality assurance of test and calibration results by, amongst other means, the regular use of CRMs and/or internal quality control using secondary reference materials and by participation in inter-laboratory comparisons or proficiency testing programmes.

Methods

Validated food analysis methods are used for compliance with food legislation in the internal EU market and global trade. They serve likewise to detect fraud, to test for the authenticity of specifically labelled food products and to monitor specific substances for exposure assessment (e.g. EU pesticides programme).

Due to the demand for reliable and comparable methods, performance requirements have been established at a national and international level for implementation of official methods, e.g. by European legislation, by the CEN or the Association of the Analytical Communities (AOAC) International, and worldwide by Codex Alimentarius (CAC). Thus any method proposed to be used for official purposes must be validated in a collaborative trial study, resulting in defined method performance characteristics [4]. The framework for the design and conduct of such collaborative trial studies, as well as the statistical evaluation, are also defined in appropriate protocols [5]. Any method that has been successfully validated according to these protocols can be recognised as an official method for use in legal cases or for international trade control purposes.

Food methods validated by a collaborative trial study and those validated using the single-laboratory approach have been adopted as national and international standards by, e.g. CEN, International Organisation for Standardisation (ISO), AOAC International and by the Joint FAO/WHO Codex Alimentarius Food Standards Programme. A number of EN Standards developed by CEN relate to the organisation of controls. It is however important to keep in mind that, in addition to the method performance criteria, economical and prevention strategy aspects are also important in method development. Demands for fast and efficient procedures and the possibility of automation have lead to the development and validation of rapid screening methods, in addition to the well-established official confirmatory methods.

Reference materials

FF-RMs can be CRMs, proficiency test materials (PTMs), test materials and other standard materials. CRMs are used for in-house verification of an externally fully validated method, as is required by ISO 17025, as well as for the assessment of the analytical recovery of a method [6].

It is essential that FF-RMs are as similar as possible to the "real" samples as also holds true for other CRMs, e.g. environmental samples [7]. This is often not possible due to very complex and delicate food and feed matrices especially as CRMs must show stability over a certain range of time. In food safety and quality, CRMs play an important role in the analysis of e.g. food microbial contamination, food contaminants such as mycotoxins, dioxins, polychlorinated biphenyls (PCBs) and food residues such as veterinary drug residues (hormones, antibiotics) and pesticides. CRMs for food quality control are important for analysis of food constituents such as fat, sugar and protein content or of typical indicators for the food origin (e.g. stable isotopes in wine). In addition, FF-RMs are used in proficiency testing, although most of this testing is done with non-certified assay materials (PTMs).

An additional and extremely useful tool can be the availability of FF-RMs in order to calibrate the analytical results achieved (e.g. in monitoring studies). Although there is sometimes the opinion that this is an expensive solution, FF-RMs can save time, add to the reliability of validation data and thus prove an enormous economic value.

The use of reference materials, when fulfilling requirements concerning matrix and analyte concentration provides a link to a stated reference. This traceability is the property of a result related to the international standard (SI) or a stated reference. A CRM is also a tool in determining the selectivity and specificity of a method.

When an FF-RM and especially a CRM is desired for validation of a measurement procedure, such material is not always easily available. This is especially the case for CRMs with naturally incurred contaminants or residues. Unfortunately these are still rare and often not matrix-matched. Examples of urgent areas of need of such FF-RMs are given in the following section.

Also, the CRM needs to have the certified component at the required level of the measurement, with a stated uncertainty at the level of intended use. A minimum portion may be recommended for the compatible CRM.

CRMs and method validation—selected examples in food safety and quality

Persistent chlorinated compounds

The occurrence of high concentrations of PCBs and dioxins in food and feed in Belgium a few years ago brought about a sudden drop in confidence in food safety. More extensive monitoring activities concerning the levels of PCBs and dioxins in food started immediately after the public announcement in individual Member States. This revealed a strong need for fast analytical methods for screening purposes. It was also found that Member States' enforcement laboratories applied varying analytical methods, which indicated the need for a comparison of results from different laboratories [8]. There are some CEN standards available that offer a number of strategies for extraction, clean-up and quantitative analysis of PCBs [9]. However, these methods do not only determine PCBs and are rather time-consuming. For this reason a simplified method based on gas chromatography with mass spectrometric detection (GC-MS) for the determination of the specific PCBs was developed and validated in-house on various food and feed matrices [10]. In this single-laboratory method validation study both spiked food and feed samples and CRMs were used, namely BCR 349 (cod liver oil) and BCR 450 (milk powder), both certified for the content of PCB congeners. The recoveries of the individual PCB congeners varied from 88 to 107%, indicating good correlation with internationally recognised criteria for the performance characteristics of analytical methods. It must be noted that one of the CRMs applied was only a contaminated oil, while most contaminated food products in the study needed special treatment or fat extraction from the matrix prior to measurement. It is obvious that the sample treatment or extraction is a major contributor to the estimation of uncertainty. By avoiding the extraction step, the method uncertainty of the result may be substantially lower than in reality.

The GC-MS technique was applied in Member States as much as possible but capacity for large numbers of samples was limited during this calamity and rather high in costs. Therefore the need for a rapid screening method to detect PCBs and dioxins became apparent. In order to evaluate the potential of an immunoassay-based method for rapid screening of PCBs in food and feed, an in-house validation study was carried out [11]. Again CRMs (CRM BCR 349 Cod liver oil and Mackerel oil CRM BCR 350) were used to ensure the traceability of the measurement results. The immunoassay incorporates a rapid sample-processing protocol. The latter has been optimised to detect concentrations of PCBs in animal fat at or above 200 ng/g. This immunoassay has shown to be suitable for the rapid determination of PCBs with a high sample throughput and minimal hazardous solvent waste. Due to the nature of both CRMs (oils), several extraction steps in the sample treatment for food and feed were not made, which certainly leads to a lower measurement uncertainty than expected for a real and complex food or feed matrix.

It is interesting to note that there was a significant increase in the sales of PCB reference materials available from the Institute for Reference Materials and Measurements (IRMM) of the European Commission's Directorate General Joint Research Centre (JRC). The IRMM is responsible for the storage of all BCR CRMs and is also producing in-house food and feed-related CRMs as well as other materials. The sales of CRM BCR 349 (Cod liver oil) increased more than 10 times in the middle of 1999 (just after the discovery of the Belgian dioxin crisis) in comparison to 1998. In order to prepare a more realistic CRM, IRMM prepared, immediately after the announcement of the Belgian crisis, PCB-contaminated pork fat samples and PCB proficiency testing calibration standards for the quality control of data produced by the analysing laboratories [12, 13]. In conclusion, it can be stated that different analytical methodologies can be applied to the determination of PCBs, provided certain quality criteria are fulfilled and CRMs have shown to be a very useful tool in assessing these methods.

Cocoa butter equivalents in chocolate

The new European Chocolate Directive [14] allows the addition of up to 5% of vegetable fats other than cocoa butter (CB), the so-called cocoa butter equivalents (CBEs), in chocolate. CBEs resemble the chemical composition and physical properties of CB very closely, making them therefore extremely difficult to quantify and even in some cases to detect (especially at very low levels). There is a perceived need within official control laboratories for reliable analytical methods for the quantification (around the 5% level) of CBEs in chocolate, as Member States' laws and administrative provisions need to comply with the new Chocolate Directive before August 2003. All proposed analytical methods have been evaluated by the JRC in collaboration with EU expert laboratories [15]. The performance of several methods has been compared and a final method based on the analysis of the main components, triglycerides, has been proposed for further validation.

A cocoa butter CRM has been prepared in the course of this project in order to facilitate the work of the analytical chemist [16]. The CRM IRMM 801 aims to ensure a high comparability of the analytical results achieved. It was used as a calibrant for the establishment of a standardised database containing data from more than 74 different CBs and 94 CBEs. The latter resulted in the application of a simple equation by testing laboratories detecting CBEs in mixture with CB and in plain chocolate. For two methods, based on gas–liquid chromatography, standardised method descriptions have been prepared and both methods were recently validated in a collaborative study [17, 18]. One method has been tailored to detect CBEs in CB and confectionery products down to a level of 0.4% related to the final product, chocolate (assumed fat content of chocolate 20%), thus limiting the false-negative or false-positive results. The other method was aimed at the determination of CBEs at the 5% level in chocolate. In both validation studies, the CB-CRM IRMM 801 was applied as a calibrant.

In conclusion, having today the two validated methods together with the cocoa butter CRM at hand, the implementation of the EU Chocolate Directive has been made feasible, at least for the assessment of the amount of CBEs in plain chocolate.

Peanut allergens in food products

The European Commission has recognised the problem of food allergens and has recently made a proposal to amend the European Food Labelling Directive that imposes that all ingredients intentionally added to food products must be included on the label [19].

However, correct labelling requires the knowledge of the concentration of the allergens and their behaviour during processing. In addition, the assessment of compliance with labelling requires suitable analytical procedures, which should be validated either by an internationally accepted in-house testing protocol or by a collaborative study.

For this reason, the CEN has recently established a new working group on food allergens in order to standardise the analytical methodology available so far (CEN TC 275 WG 12). The working group concluded that there are no collaboratively studied methods currently available for the analysis of allergens in the low ppm range.

The detection, and especially the quantification, of allergens in processed food products can be very difficult, as they are often present in trace amounts only, or are masked by the food matrix. It was shown that a level of 100 μg of peanut proteins can already trigger a mild reaction in a peanut-allergic person. This could be caused by the consumption of 100 g chocolate or biscuits containing 1 mg/kg peanuts.

The IRMM has recently started the validation of presently available methods (mostly based on enzyme-linked immunosorbent assays) in a collaborative trial study with real food samples such as cookies and dark chocolate containing peanuts in minute amounts (1–20 mg/kg). These methods have been already been validated in-house [20].

The capability to detect any unintentional contamination of food products that usually do not contain peanuts is especially important for peanut-allergic patients. Detection limits for peanut allergens probably need to be 1 mg/kg or even lower.

Validated methods are just as important as the availability of reference materials for allergens [21]. As peanut allergy is highly prevalent and peanut products may enter into the production of various food matrices, e.g. chocolate, ice cream, biscuits and breakfast cereals, it is essential to have a peanut reference material both for research and routine analysis. Peanuts available in the food sector are derived from various sources, such as peanut varieties/types from different geographical origins, and are treated by various technological processes, such as dry and oil roasting at various temperatures for various times.

The IRMM has launched the production of a basic peanut reference material using the most commonly used peanut varieties for food production. This material may be used for spiking food matrices for further method development, validation and proficiency testing, for clinical tests and in vitro assays. This future peanut reference standard will take into account specific demands from the food industry and respect various technological conditions as mentioned.

Future CRMs for food and feed analysis—outlook

For food safety and quality control, a number of other CRMs would be very much welcomed. Many requests are made for matrix matched CRMs and PTMs. Especially official food and feed control laboratories and moreover the Community and National Reference Laboratories are dependent on fit-for-purpose validated analytical methods and real matrix reference materials.

In order to match with method validation studies recently carried out, amongst the ones high on the list are CRMs for smoke flavourings, both for toxic substances (e.g. polycyclic aromatic hydrocarbons) as well as their compositional characterisation, tissue of the central nervous system (e.g. brain in meat products), pesticides in food and feed for the EU pesticides monitoring programme, veterinary drug residues and hormones in animal-derived tissues, food allergens, and especially microbial contaminated food and feed matrices. Other highly desired CRM's are those containing mycotoxins in food at the latest legislative levels, e.g. patulins in apple products and baby food, fusarium toxins in cereals, acrylamide in food or products from organic farming for proof of authenticity.

Conclusions

Food and feed CRMs, when available and meeting the necessary conditions, are an extremely useful tool in validation of measurements for both calibration and traceability. The homogeneity, stability and assignment of a certified reference value with a CRM (with associated stated uncertainty) enhance the reliability of the method performance assessment. This has been recognised by international standards and by European legislators. Therefore the appropriate use of CRMs and validated methods has a strong impact on legal action levels. There is an immense need for many more FF-RMs than are available today. However, it should be clearly stated that validated methods and CRMs do not automatically guarantee the quality of the analytical results.

References

1. European Commission (2003) Proposal for a Regulation of the European Parliament and of the Council on official feed and food controls. EU, Brussels
2. European Commission (1993) Council directive 93/99/EEC of 29 October 1993 on the subject of additional measures concerning the official control of foodstuffs. Off J L 290, 24/11/1993, pp 14–17
3. International Organisation for Standard ISO/IEC 17025 (1999) General requirements for the competence of testing and calibration laboratories. ISO, Geneva
4. European Committee for Standardisation, CEN (1999) Food analysis—biotoxins—criteria of analytical methods for mycotoxins, Report Reference no. CR 13505:1999E
5. Horwitz W (1995) Pure Appl Chem 67:331–343
6. Barwick V, Burke S, Lawn R, Roper P, Walker R (2001) Applications of reference materials in analytical chemistry. LGC, Teddington, UK. ISBN 0-85404-448-5
7. Fajgelj A, Parkany M (1996) The use of matrix reference materials in environmental analytical processes. RSC, UK. ISBN 0-85404-739-5
8. von Holst C, Mueller A (2001) Fres J Anal Chem 371:994–1000
9. European Committee for Standardisation CEN (1997) EN 1528-1, 1528-2, 1528-3 and 1528-4. Beuth Verlag, Berlin
10. von Holst C, Mueller A, Bjoerklund E, Anklam E (2001) Eur Food Res Technol 213:154–160
11. Jaborek-Hugo S, von Holst C, Allen R, Stewart T, Willey J, Anklam E (2001) Food Add Contam 18:121–127
12. Bester K, De Vos P, Le Guern L, Harbeck S, Kramer G (2000) The certification of PCBs in pig fat—IRMM 444, 445 and 446, EU-Report EUR 19600 EN
13. Bester K, De Vos P, Le Guern L, Harbeck S, Kramer G (2001) Chemosphere 44:529
14. European Commission (2000) Directive 2000/36/EC of the European Parliament and the Council of 23 June 2000 relating to cocoa and chocolate products intended for human consumption. Off J L197, pp 19–25

15. Lipp M, Simoneau C, Ulberth F, Anklam E (1999) Possibilities and limits to detect cocoa butter equivalents in mixture with cocoa butter—devising analytical methods for the determination of cocoa butter and other vegetable fats in chocolate with the aim of checking compliance with existing Community labelling. EU Report EUR 18992 EN
16. Koeber R, Buchgraber M, Ulberth F, Bacarolo R, Bernreuther A, Schimmel H, Anklam E, Pauwels J (2003) The certification of the content of five triglycerides in cocoa butter. EU Report 20781 EN
17. Buchgraber M, Anklam E (2003) Validation of a method for the detection of cocoa butter equivalent in cocoa butter and plain chocolate. Report on a validation study. 2003 EUR 20685 EN
18. Buchgraber M, Anklam E (2003) Validation of a method for the quantification of cocoa butter equivalents in cocoa butter and plain chocolate, EU-Report 20777 EN
19. European Commission (2000) Food labelling directive. Off J L109, 6.5.2000
20. Poms RE, Lisi C, Summa C, Stroka J, Anklam E (2003) In-house validation of commercially available ELISA test kits for peanut allergens in food. EU-Report 20767 EN
21. Poms RE, Ulberth F, Klein CL, Anklam E (2003) GIT Lab J Eur 3:132–135

Accred Qual Assur (1999) 4:352–356

D. Brynn Hibbert

Method validation of modern analytical techniques

Abstract Validation of analytical methods of well-characterised systems, such as are found in the pharmaceutical industry, is based on a series of experimental procedures to establish: selectivity, sensitivity, repeatability, reproducibility, linearity of calibration, detection limit and limit of determination, and robustness. It is argued that these headings become more difficult to apply as the complexity of the analysis increases. Analysis of environmental samples is given as an example. Modern methods of analysis that use arrays of sensors challenge validation. The output may be a classification rather than a concentration of analyte, it may have been established by imprecise methods such as the responses of human taste panels, and the state space of possible responses is too large to cover in any experimental-design procedure. Moreover the process of data analysis may be done by non-linear methods such as neural networks. Validation of systems that rely on computer software is well established. The combination of software validation with validation of the analytical responses of the hardware is the challenge for the analytical chemist. As with validation of automated equipment such as programmable logic controllers in the synthesis of pharmaceuticals, method developers may need to concentrate on the process of validation, as well as the minutiae of what is done.

Key words Method validation · Sensors · Electronic nose

Introduction

Method validation is an established process which is:
"the provision of documentary evidence that a system fulfils its pre-defined specification" [1], or:
"the process of proving that an analytical method is acceptable for its intended purpose" [2].

EURACHEM has offered a more long-winded definition that includes: "checks... to ensure that the performance characteristics of the method are understood and demonstrate that the method is scientifically sound under the conditions under which it is to be applied" [3].

In theory any method used by an accredited laboratory will have some validation documentation, either provided by the laboratory or because it is using an official method that has been validated. In the latter case the laboratory must provide evidence that it can carry out this method competently (verification). What is required to validate a method is clear in the case of regulated pharmaceutical laboratories, and although some of the accepted validation statistics may be questioned (for example $r>0.999$ for calibration linearity [4]), the process may be said to produce the desired result, i.e. methods employed to analyse pharmaceuticals are fit for purpose.

The contention of this paper is that apart from this example, methods used for other purposes are often not well validated, and as increasing use is made of multi-sensor methods involving intelligent data analysis, there may not be even a prospect of validation in the traditional sense.

Method validation of analytical methods for pharmaceuticals

In the context of this paper, there are three aspects of the validation of analytical methods used in the pharmaceutical industry that are of note. First, the product is of high cost, and the industry is very profitable. Secondly, it is one of the most regulated industries, often with separate agencies overseeing its activities. (In Australia, this is the Therapeutic Goods Administration). Thirdly, although there is an increasing interest to determine the nature and concentrations of possible excipients and breakdown products arising from the manufacture and storage of pharmaceuticals, the systems for analysis are essentially well known, both in terms of identification and concentration. Therefore the industry is required to validate by law, they have the money to do it, and the system is tractable. It is the contention of this paper that if validation is possible for any method, validation for the pharmaceutical industry should be the most straightforward. This does not, of course, underestimate the time and cost that goes into a full-method validation. Guidelines for validation of pharmaceutical methods have been published by a number of bodies [2, 3, 5].

Validation is a logical process that is conducted in parallel with method development. If a method that is being developed will ever be used in earnest, then it must be validated. The data obtained during method development may be used towards validation, and the outcome of validation may force changes in the method (which must then be re-validated). This iterative procedure is known as the 'development/validation cycle'. Methods that are submitted for regulatory approval must show evidence of validation of the attributes summarised in Table 1. Before the method is developed, in consultation with the end-user, minimum acceptance criteria must be determined for each of the aspects of the method given in Table 1. The validation report will then document how the method has met the criteria.

The methods may be done in blocks, with the later validation steps, such as robustness being the 'icing on the cake' of the method. In large companies, decisions will be made at different stages to continue work on the method and validation, or to cease development. Some aspects of the validation process are discussed below.

Specificity

The number and concentrations of impurities represents one of the few, possibly indeterminate quantities in pharmaceutical method validation. However, careful research into all stages of manufacture, storage and use can reveal the range of compounds that are not the active product. Modern advances in liquid chromatography with diode array or mass spectrometric detection have aided the identification of impurities. Peak purity of the target analyte is assessed after the product has been deliberately 'stressed' by exposing it to high temperature, high humidity and high-light levels. For bulk pharmaceuticals exposure to acid, base and oxidising agents may also be studied. When validating impurity methods, the resolution of the impurity peaks, among themselves and from the active ingredient peak, is of importance.

Table 1 Validation of an analytical method, showing tasks and typical acceptance criteria for the analyte

Validation studies	Tasks for analyte method
Specificity	Analyte + placebo, synthesis intermediates, excipients, degradation impurities; versus pure analyte. Peak purity and resolution assessed. Resolution >1.5
Calibration linearity	Six levels each with three replicates from 50% to 150% of target concentration. $r>0.999$
Accuracy	Recovery of certified reference material. ±2% at concentrations of 80% to 120% of target concentration.
Precision	Instrument repeatability – ten replicate injections. RSD<1% Intra-assay precision. Multiple analysis of aliquots of a sample during one day. RSD<2% Intermediate precision. Multiple analysts, instruments, days in one laboratory. Reproducibility by inter-laboratory studies to detect bias.
Calibration range	Determine from accuracy and linearity studies.
Detection limit	Only for impurity methods. S/N=3 from blank studies or calibration.
Quantitation limit	Only for impurity methods. Concentration to give a defined RSD (e.g. 10%), or S/N=10.
Robustness	Experimental design establishing the changing critical parameters (e.g. mobile phase composition, column temperature).

Calibration linearity

Although non-linear calibration is widely available, it is still considered mandatory that across the range of likely use (80%–100%, or more widely 50%–150% of target concentration) the calibration line to establish the relationship between the measured quantity (peak height, peak area) and the concentration of analyte should be linear. A commonly used criterion is that the correlation coefficient be greater than 0.999. The square of the correlation coefficient (sometimes known as the coefficient of determination) is the fraction of the variance in the dependent variable, explained by the independent variable. As such it can be used for assessing the validity of a linear model, but it is not a good measure of scatter about a line for which linearity is established. Mulholland and Hibbert have shown that for slightly curved lines, even with $r=0.999$, when the line was used for calibration, errors in the predicted concentration of 70% could be observed at the lower end of the range [4]. If such linearity is established, one-point or three-point calibration may then be used in practice. Mulholland and Hibbert [4] recommend inspection of residuals, and Green [2] the use of the response factor $= (y_I-a)/x$ for a linear relationship $y=a+bx$. Each of these methods highlights small deviations from linearity that are not apparent from the simple plot of y against x. The use of the calibration line to establish a detection limit has been promulgated by ISO [6] as a more realistic method than the generally used 3 times the standard deviation of the baseline.

Precision

Precision is usually measured as the standard deviation of a set of data. Of importance is which set of data, taken under what circumstances. The more variability that is allowed in a set of experiments, the greater the variance that will result. Thus if ten aliquots of a homogeneous sample are injected into a single chromatograph by a single analyst one after the other, the *repeatability*, (as the standard deviation of the ten peak heights is known) would be expected to be no more than 1% of the average peak height. Unfortunately in papers submitted to journals, this is often the only measure of a method's precision. Intra-assay, or intra-laboratory precision measures the effects of different analysts, or repeated sample preparation. It is the standard deviation (usually reported as a relative standard deviation, i.e. the standard deviation divided by the mean, also known as the coefficient of variation) of repeated analyses of aliquots of a single sample on 1 day in the laboratory. As part of a ruggedness study an experimental design may be undertaken to vary a number of possible parameters with a view to determine which are important to the analytical method.

Highly fractionated designs are used (Taguchi or other orthogonal designs) which establish main effects only with the minimum number of experiments. Finally, as part of assessment by regulatory bodies, a laboratory may participate in a proficiency study in which their method will be used to analyse two unknown samples containing a stated analyte. The inter-laboratory precision is usually at least twice that of the intra-laboratory precision, and that after removal of outliers. It is now customary to use a robust statistic such as a robust z score, which is defined as

$$z = \frac{(x - median(x))}{normIQR}.$$

normIQR is the normalised interquartile range, i.e. the range which encompasses the middle 50% of a set of data multiplied by 0.7413. Essentially the robust z score reports how many standard deviations a sample is from the middle of the data set. Outliers are then identified as having $z>3$. Such interlaboratory comparisons are a useful way of establishing conformity among a group of laboratories, but it must be stressed that unless certified samples are used, it does not establish accuracy. As De Bièvre has observed, the outlier has sometimes been shown to be the only laboratory near to the correct answer.

An experimental determination of reproducibility should also coincide with a theoretical calculation based on the known, or estimated, precision of the individual steps of the method. The so-called 'bottom up' method as part of an uncertainty audit, is included in the principles of valid analytical measurement (VAM) [7–9].

Environmental analysis

Before considering the modern techniques that will prove almost impossible to validate in the sense described above, I shall comment on a sector of analytical chemistry that, while superficially similar to the pharmaceutical industry, by virtue of less regulation and more complex samples, already provides examples of difficulty of validation.

At a recent conference on analytical chemistry held by the Royal Australian Chemical Institute [10], an environmental consultant, bemoaned the presence of analytical laboratories which, while being fully registered with the appropriate body (in Australia the National Association of Testing Authorities – NATA) did not consistently provide results that were 'fit for purpose'. Costs were being cut to the point that, in the opinion of the speaker, the results were almost meaningless. What is the difference between an analytical method that is designed to determine the correct concentration of an active ingredient of a pharmaceutical product and a method that will determine if a sample of soil has greater than the action limit of heavy metals? First the system in environmental analysis is less well defined and could be drawn from a great number of possibilities. As a consequence the matrix is not completely defined. Secondly, even the approximate concentration of the analyte may not be known. There is no target concentration, the amount of a given heavy metal could vary between zero and many thousands of parts per million. Sampling and sample pretreatment are also crucial for obtaining results that are meaningful. In terms of validation it is possible to comment on the effect of these differences.

Specificity

In terms of the sample, specificity of the method is difficult to establish. It may be that the clean up procedure allows the final sample to be analysed free of interferences, but the process of obtaining such a laboratory sample from the original material in the environment has a great number of uncontrollable variables. Obvious interferents may be known and procedures adopted to avoid them. An example is the presence of high levels of sodium chloride in sea water samples, which proves difficult for atomic spectroscopy methods.

Calibration linearity

The linearity of calibration should be established. However in practice laboratories using ICPOES or ICPMS tend to calibrate over a very wide range, often using single-point calibration in the vicinity of the concentration of that of the sample. The problem arises with a batch of samples that show a wide variation in concentrations of analyte. Proper attention to the calibration protocols will require a number of ranges to be validated. Again, in practice, this may not be adhered to because of pressure of the number of samples to be processed.

Accuracy

The nature of the samples means that a CRM to establish accuracy in a proper metrological way may not exist. Usually, recovery studies will be done with spiked samples. There is concern about the purity of standards used here, and also about the speciation of redox active species such as transition metal elements. Standard addition is a method that extrapolates beyond the calibration range, and is therefore prone to high uncertainty. The low levels of analyte makes it difficult to establish a 'true' concentration.

Precision

Once a method is established, precision may be determined by suitable replicated experiments. However it is in inter-laboratory trails that the problems with environmental methods often show up. It is accepted that for trace analysis RSD values of tens of percent are likely. In studies conducted in Western Australia on pesticide residues in bovine fat RSD values for dieldrin were 12% and for diazonium were 28%. It is typical to see a quarter of the laboratories in such a trial producing values that could be termed outliers. In the previously mentioned study, 5 laboratories out of 26 had $z > 3$ for aldrin. In a parallel study RSD values for petroleum in dichloromethane and water were 40% and 25%, respectively. The conclusions of these studies was that there was poor comparability because of the different methods used, that accreditation apparently made no difference to the quality of results, and that a lack of understanding of definitions of the quantities to be analysed (for example 'gasoline range organics') caused non-method errors. In relation to methods, this is contrary to the conclusion of van Nevel et al. who asserted that the results of the IMEP round of analyses of trace elements in natural and synthetic waters showed no dependence on method [11]. If properly validated methods do yield comparable results, then one conclusion from the range of studies around the world is that many environmental methods are not validated. It may be that validated methods are indeed used, but not for exactly the systems for which they were validated.

Limits of detection and determination

Much is made of detection limits in environmental analysis. Much of the modern concern about chemicals in the environment stems from the ability of the analytical chemist to analyse ever lower concentrations. Less attention is given to a limit of determination. Usually the RSD is the best possible for the given method, and because intra-laboratory precision, or even simple repeatability, is quoted, this is usually accepted.

Array sensors

The ultimate goal of analytical chemistry is not to analyse chemicals per se but to solve problems couched in the language of society. "Can we drink the water?" "Can a children's playground be built on the waste site?" "Does the patient have an over active thyroid?" [12]. Society also wants the answers to questions like these as quickly as possible and as cheaply as possible. These pressures have given a boost to the development of portable sensors that yield answers that are intelligible to the untrained user. Computer control and in situ data analysis have made great strides, and it may be argued that the development of suitable chemical sensors is the limiting factor in the wider use of these devices [13]. Arrays of sensors that relay on sophisticated data analysis are typified by the probably missnamed "electronic nose" [14]. Portable sensors have been developed for monitoring atmospheric pollution, foods, wines and other beverages, odours from factories, abattoirs and sewage plants [15].

Analytical methods using arrays of sensors

The transduction mechanisms of these sensors are based on the conduction of semiconductors such as tin oxide [16], or polymers such as polypyrrole [17]. More sensitive are sensors that 'weigh' impinging molecules [18] and more sensitive still is the biological nose. Recently there has been a renewal of interest in optical sensors incorporating fluorescent molecules [19]. Typically a device will have 3 to 30 sensors, the output of each being a voltage. This may be measured at the steady state, or the time development of the voltages may be monitored. Humidity and temperature control is important for many sensors.

The array may target a particular volatile chemical, or it may be calibrated against less exact standards, for example, the subjective scores given by a taste-testing panel.

Data treatment and analysis

The methods of data analysis depend on the nature of the final output. If the problem is one of classification, a number of multivariate classifiers are available such as those based on principal components analysis (SIMCA), cluster analysis and discriminant analysis, or non-linear artificial neural networks. If the required output is a continuous variable, such as a concentration, then partial least squares regression or principal component regression are often used [20].

Using such an array for, for example, the provenance of a red wine claiming to be Chianti, requires calibration with a number of samples that cover the range of genuine Chianti and all the ersatz versions that may be encountered. An unknown sample is presented to the array and the output (usually a number of voltage responses) is fed into the data analysis software, which in due course, returns the answer "The wine is Chianti" or "The wine is not Chianti". With more information during calibration, the response may offer further advice on the origin of the sample, or it may determine how near a true Chianti, a forgery may be.

Aspects of method validation

Given that an array sensor is being developed, how may they be validated? I know of no system that has undergone validation for accreditation purposes. Even compared with environmental analysis, these sensors present difficulties that need to be resolved, before a consensus on suitable validation protocols is reached. Gardner and Bartlett have considered the problem of how to define the performance of electronic noses using standard odour mixtures [21]. They propose two indicators of performance, the range of different odours that may be detected by the array, and the ability to discriminate between similar odours (the resolving power).

Accuracy and precision

Precision and accuracy need to be redefined for classification problems. The accuracy is the percentage of correct classifications, when a series of unknowns are presented to the calibrated instrument. For systems in which there is a difference in the consequence of false positives or false negatives (for example medical diagnosis of an infectious disease) the test of accuracy has to be set up to reflect this difference. The repeatability of a classification system may also be determined by presenting the same sample a number of times to it, and recording the percentage of correct classifications. The statistics of discrete events can be applied to the results to yield probability levels. Just as digital signal encoding has advantages over analogue methods, a discrete classifier may be expected to produce a higher percentage of correct answers.

Calibration

Headings such as 'calibration linearity' have no direct meaning for multivariate methods that do not have a single 'x' value. Indeed, multivariate regression methods pose the calibration problem with the 'Y block' being the target variable (e.g. concentration) and the 'X block', the matrix of measurands (e.g. voltage outputs of the array of sensors), this being the inverse of a traditional analytical calibration graph [22]. Multivariate calibration, because of the complexities of models and the number of parameters, must be shown to be valid both in terms of the calibration model (i.e. how well the model fits the standards used in calibration), and its prediction ability. Two methods to establish the prediction ability of a method are popular. First, if there are sufficient calibration sets of data, a tranche (say 10%) is set aside to be used to validate the model. The validation set is presented to the model and the success of the prediction is assessed. Alternatively, a leave-one-out procedure is adopted, known as cross validation, in which the model is established with n–1 data points and the n th data set is predicted by the model. The n th set is returned and a different set is removed, the model established and the removed set predicted. This continues until all of the data sets have been removed and predicted. Cross validation does not work if the calibration set has been determined by a minimal experimental design, because each set is vital to the statistical integrity of the whole.

Robustness

The most difficult aspect of validation to establish for an array sensor is the robustness, particularly in terms of the variation in the data sets that may be presented to the instrument. In the example used above, while a number of genuine Chianti vintages may be sourced and presented to the device, a similar library of fake Chiantis may not be available. The calibration depends greatly on the set of cases used, and this also impinges on the robustness of the model. If data from a human taste-testing panel is used to establish the model, then a number of different groups of tasters should be employed to establish the sensitivity of the model to the vagaries of the human panels.

Validation of software

Most array sensors rely on computerised algorithms to build the calibration model. In-house software or customised commercial products are often used. The need to establish compliance of the software is evident, both in terms of its own specifications and in use with the instrument. Software should be certified to ISO 9001 under the TickIT scheme [23–25], and in addition to basic performance requirements the following should be addressed in the user requirements specification:
- Alarm, event and message handling
- Audit, report and archiving requirements
- Transfer of data via networks
- Display and user-interface specification
- Maintenance, security and diagnostics
- System failure and recovery requirements to protect the integrity of data
- Upgrading and future requirements
- Year 2000 compliance.

Before acceptance, therefore, the above user requirements should be thoroughly tested according to an agreed protocol, and further tests must be undertaken on the use of the software with the instrument (field testing).

Discussion

The validation of analytical methods that are based on arrays of sensors cannot rely on the traditional pharmaceutical approach. Many of the validation headings have no equivalent, and important steps, such as validation of software, are not explicitly included. The pharmaceutical industry does, however, have a model for validation that could be used in modern analytical chemistry, namely the validation of programmable logic controllers (PLC) [26] and supervisory control and data acquisition (SCADA) [27] systems. This was the subject of a recent issue of the journal *Measurement and Control* [1]. Of importance is now the process of validation, which may be broken down into a number of steps (Fig. 1).

The validation life cycle of Fig. 1 shows validation to be the driver of the development process, coming before any development is commenced, and with the different steps of qualification verifying a set of specifications. The final validation report closes the cycle and is the written evidence that the system has been validated and is therefore 'fit for purpose'. Parameters such as accuracy and precision, now become requirements within one of the validation stages. Installation qualification ensures that all of the components of the method are present and installed according to the design specification. Operational qualification ensures that each part of the method functions according to specification when tested against a representative set of cases, and performance qualification tests the performance in a real environment, including tests of software that were enumerated above.

Complex instruments giving advice as the output may never be validated to the extent of a simple analytical method. However, what validation will do is to set the bounds of the responses and situations in which the instrument will give fit-for-purpose output. The users of the instrument will then need to establish if their particular uses fall within the scope of the validated instrument. We look forward to a report of a fully validated electronic nose.

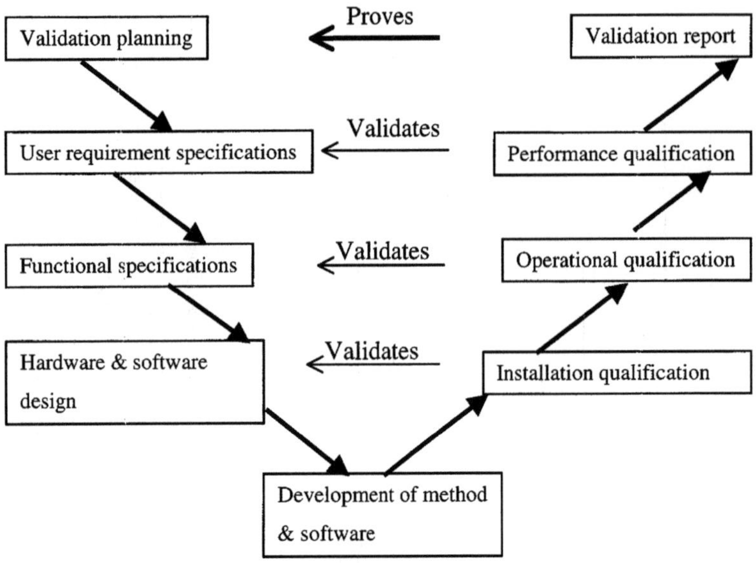

Fig. 1 The validation life cycle (adapted from [26])

Conclusion

In this paper we have reviewed the validation of analytical methods used in the pharmaceutical industry, and conclude that the methods do establish fitness for purpose. However outside this tightly regulated industry, it is not so clear that properly validated analytical methods are used. Contemplation of methods of environmental analysis suggests that the complexity of the problems and the use of methods outside the confines for which they were originally validated, may be a contributing factor to the concern currently expressed about the quality of analytical results. Modern analytical methods employing arrays of sensors, using multivariate calibration models and providing discrete output (classification), adds even more complexity to the validation problem. The validation of PLC in the pharmaceutical industry is given as an example of the validation of a highly complex system, and it is suggested that approach taken with PLC validation be adopted when validating modern analytical methods.

References

1. Wingate GAS (1998) Meas Control 31:5–10
2. Green JM (1996) Anal Chem 68:305–309A
3. EURACHEM (1997) Guidance document No. 1/WELAC Guidance document No. WGD 2: Guidance on the interpretation of the EN 45000 series of standards and ISO/IEC Guide 25. Laboratory of the Government Chemist, Teddington, UK
4. Mulholland MI, Hibbert DB (1997) J Chromatogr 762:73–82
5. Food and Drug Administration (1996) Draft guidelines on the validation of analytical procedures: methodology. Federal Register 61:9316–9319
6. ISO 11095 (1996) Linear calibration using reference materials International Organisation for Standardisation, Geneva
7. Sargent M (1996) Confidence in Analysis. Laboratory of the Government Chemist, Teddington, UK. VAM Bulletin (Autumn 1996) 3–4
8. ISO (1995) Guide to the expression of uncertainty in measurement. International Organisation for Standardisation, Geneva
9. EURACHEM (1995) Quantifying uncertainty in analytical measurement. Laboratory of the Government Chemist, Teddington, UK
10. McLeod S (ed) (1997) Proceedings of the 14th Australian Symposium on Analytical Chemistry, Adelaide July 1997. Analytical Chemistry Division, Royal Australian Chemical Institute, Melbourne
11. van Nevel L, Taylor PDP, Örnemark U, Moody JR, Heumann KG, De Bièvre P (1998) Accred Qual Assur 3:56–68
12. Mulholland MI, Hibbert DB, Haddad PR, Parslov P (1995) Chemometrics Intell Lab Syst 30:117–128
13. Haswell SJ, Walmsley AD (1991) Anal Proc London 28:115–17
14. Gardner JW, Bartlett PN (1994) Sens Actuators B 18:211–20
15. Hodgins D, Simmonds D (1995) J Autom Chem 17:179–85
16. Chiba A (1990) Chem Sens Technol 2:1 –18
17. Hierlemann A, Weimar U, Kraus G, Schweizer-Berberich M, Goepel W (1995) Sens Actuators B 26:126–34
18. Barko G, Papp B, Hlavay J (1995) Talanta 42:475–82
19. Dickinson TA, White J, Kauer JS, Walt DR (1996) Nature 382:697–700
20. Hibbert DB (1996) In: Linskens HF, Jackson JF (eds) Modern methods of plant analysis vol 19: Analysis of plant fragrance. Springer, Berlin, pp 119–140
21. Gardner JW, Bartlett PN (1996) Sens Actuators B 33:60–67
22. Martens H, Næs T (1989) Multivariate calibration. Wiley, Chichester, England
23. ISO 9001 (1994) Quality systems – Model for quality assurance in design, development, production, installation and servicing. International Organisation for Standardisation, Geneva
24. ISO (1991) Quality management and quality assurance standards – Part 3: Guidelines for the application of ISO 9001 to the development, supply and maintenance of software. International Organisation for Standardisation, Geneva
25. DISC TickIT Office (1995) The TickIT Guide: A guide to software quality management system construction and certification. Using ISO 9001:1994, British Standards Institution, Issue 3, London
26. Mitchell R, Roscoe S (1998) Meas Control 31:10–13
27. Coady PJ (1998) Meas Control 31:14–19

Presented at: CITAC International Symposium on Analytical Quality Assurance for the 21st Century, Sydney, Australia, 15–16 October 1998

D.B. Hibbert (✉)
Department of Analytical Chemistry, University of New South Wales, Sydney NSW 2052, Australia
e-mail: b.hibbert@unsw.edu.au
Tel.: +61-2-9385 4713
Fax: +61-2-9385 6141

Relevant accreditation policies and concepts in EU-, EFTA-, NAFTA- and other regions

Accred Qual Assur (1998) 3:29–32

Permanent Liaison Group between EAL (European co-operation for Accreditation of Laboratories) and EUROLAB

Validation of test methods

General principles and concepts

Foreword

EAL and Eurolab have a Permanent Liaison Group (PLG), which is a forum where EAL and Eurolab are discussing matters of mutual interest. The PLG consists of five members from each organization.

This document has been prepared in the PLG and endorsed by both organizations.

The document is intended to give general views on certain issues related to the validation of test methods and should be seen as a common understanding and position of EAL and Eurolab. In order to define and describe the activities behind the concept "Validation of test methods" more detailed guidance documents are needed. This document should be seen as a basis for such guidance documents.

Validation of test methods

Introduction

The definition used for "validation" in the ISO standard 8402 is "confirmation by examination and provision of objective evidence that the particular requirements for a specific intended use are fulfilled". This definition gives the impression of confined and well-defined (exact) operations. Test methods are normally developed for an intended application range. The reference to particular requirements must in many cases be interpreted in a flexible way as the requirements can be of general nature.

Both standardized and non-standardized test methods are covered. They can be exact or associated with large uncertainties. Even novel test methods will be considered. The validation of a test method becomes in this context a way of demonstrating that the method is fit for its intended purpose. The fitness for purpose includes an assessment and a balancing of technological possibilities, risks and costs.

There are very few papers in the open literature dealing with the general principles of test method validation. On the other hand, a lot of detailed descriptions of the validation of specific test methods are available. A brief overview of the concepts, aims and procedures in validation is given in this document.

General principles to be used in validation

In the validation process an estimate is made of the representativeness, repeatability and reproducibility of the test method. The definitions are given in annex 1.

In the validation process the ultimate aim is to secure that the test methods are good enough with respect to representativeness, reproducibility and repeatability. How much effort should be spent on validation must be decided on a case by case basis. If large economic values as well as considerable health, safety and environmental issues are involved, much more emphasis must be paid to the validation of the test methods. The frequency of use of the test method should also be considered when determining the extent of validation. The total consequences of wrong results are of course larger for methods in extensive use than for test methods used occasionally.

The validation of test methods covers to a large extent the uncertainty, repeatability and reproducibility of the test method. As the factors affecting the results and contributing most to the uncertainty change from one technical sector to another or even from one test method to another, a universal solution cannot be given. Guidance on the expression of uncertainties can be found for example in the international "Guide to the expression of uncertainty in measurement" and EAL guidance document "Expression of uncertainty in quantitative testing".

Standardized test methods should be considered validated for their intended application range and thus good enough for that purpose although their repeatability and reproducibility are not known in detail. The testing laboratory must, however, check that they apply the method correctly. For non-standardized test methods it is up to the testing laboratories to determine how far they go in defining the level of repeatability and reproducibility.

To develop a representative test method, adequate knowledge is required of the practical use of the test results and of the real service conditions of the object of the test. Based on such knowledge, the "representative" properties to be determined by the test may be identified.

The factors affecting the test results and their uncertainty may be grouped into three main categories:

Instrumental and technical factors
- sampling
- homogeneity
- test method
- equipment

Human factors

Environmental factors
– testing environment

Instrumental and technical factors are related to the constructional and functional characteristics of the test and measurement equipment, as well as to other technical operations involved in the test (e.g. sampling, preparation of samples, test object homogeneity). Their effect may be minimized and kept under control by the following provisions:

- define the equipment as precisely as necessary
- provide a clear description of the test procedure as well as the equipment operation
- establish procedures for operational control and calibration
- ensure where applicable traceability of measurements to the SI units.

Whenever practical, the above provisions should be included in the description of the test method. References to internal procedures or applicable standards should be included.

Human factors are related to the competence of the staff and may be controlled through:

- education/basic knowledge
- on job training/practical experience.

The qualification required for the personnel employed for a given test may be specified in the test method or reference can be made to the applicable internal procedures.

Environmental factors are associated to the environment where the test is performed. Among others the effect of the following parameters must be assessed and properly controlled:

- atmospheric conditions (temperature, pressure, humidity)
- pollution/contamination
- other environmental characteristics (e.g. EMC).

The effect of the above parameters should be described in the test method or reference to other applicable documents should be made. However, for new test methods this information is often not available. In some cases the data base for method validation is so large that statistical methods should be applied.

The validation process must consider the expected or required uncertainty of the test results and their intended use.

Critical threshold values (e.g. in health and environment) cannot generally be technically justified with a small uncertainty. However, if a legal limit is set, there must be test methods suited for the purpose. Reference is made to a recent ILAC Guide.

The required depth of the validation process depends also on the maturity of the test method and the prevalence of its use. One can distinguish between the following categories:

- novel methods
- methods used by several laboratories
- modification of established methods
- standardized methods.

The ways, in which the validation is performed in the different cases, need not be clearly differentiated. If the fitness for purpose concept is maintained, it is often possible to validate at reasonable cost but with a higher degree of uncertainty.

The novel methods are first developed in one single laboratory, often on the basis of a special request from a customer or on ideas created in the laboratory. That customer cannot pay for a wide range validation nor can the laboratory itself. The aim of the validation of test methods must always be to demonstrate that the method is fit for the intended purpose and that the results have an acceptable uncertainty. It is important that the rules of validation of test methods do not prevent the natural technological development from taking place. The laboratory does not expect (although it does want) outside financial help for validation of novel methods and in many cases tries to protect its new development from going to its competitors or from becoming generally available to all.

When a certain number of laboratories work in the same area, cooperation and inter-laboratory comparisons can be arranged. The coordination of such activities is an extra economic burden. In order to speed up the process, external financing is needed.

The testing laboratories need to update their existing test methods. The flexible scope of accreditation as agreed between EAL and Eurolab was also intended to allow modifications to be made to accepted (accreditation covered) test methods. This requires validation procedures applicable to method modifications. It is up to the laboratories to describe their procedures for validating modified test methods.

The most thorough validation procedure is required for test method standardization purposes. The work needed is considerable and covers proficiency testing, the determination of factors affecting the uncertainty, measuring range, etc. The financial burden cannot be laid on the laboratories but on the standardization organizations. Standardized test methods must be considered sufficiently validated for their intended application ranges. If they are not, they should be withdrawn.

The validation of test methods consists of two interrelated steps:

(i) suitability of the test to solve the problem (customer needs)
(ii) demonstration of the technical capability of the test method within the specified test range

i.e. measuring the right properties with a sufficiently reliable method.

The suitability or representativeness of a test method is in many cases an attribute which is difficult to define especially for tests related to product acceptance. The test methods must be such that the results obtained correlate with the performance characteristics and operational experience of the product.

Validation procedure

Both testing laboratories and accreditation bodies are looking for procedures and guidelines for planning and controlling the test method validation process. However, the discussion above has clearly indicated that one single procedure cannot be developed. Consequently, a palette of different choices of validation techniques has to be developed. How detailed the validation will be, depends on the circumstances (needs, costs, possibilities, risks, etc.).

The validation of the test methods is, of course, of interest also to the accreditation bodies. The principle to be applied should be that the laboratory describes the way it is validating the test methods and the accreditation body should make the judgement if the procedure used is acceptable in that case. The different validation possibilities are built up around

- utilization of calibration
- intercomparisons including the use of reference materials and reference methods
- well qualified staff and their professional judgement
- simulation and modelling
- other approaches.

Method validation is often based on the combined use of validation procedures. The validation used can be "direct" or comparative. The selection of the validation procedures should also be justified on a cost-benefit basis as long as the fitness-for-purpose is maintained. Focusing the effort on the most critical factors affecting the test method will lead to a different solution for the validation of "exact" physical and chemical test methods as compared to that for product or subjective testing. For example, in the validation of ergonomics and sensory test methods not all possibilities are applicable.

As said above different validation procedures may be followed, their effectiveness and applicability depending on the type of test considered. They can be characterized as "scientific" or "comparative":

Scientific approach

In the scientific approach the assessment of the representativeness, repeatability and reproducibility of the method is performed with reference to the different constitutive elements and features. Evidence should describe the representativeness of the selected properties and the associated uncertainty. This can be based on information published in the scientific and technical literature or on ad hoc investigations performed by the laboratory developing the method. The laboratory shall demonstrate that relevant influencing factors (instrumental and technical, human, environmental) have been analyzed and that they are under control within the uncertainty associated with themethod.

Comparative approach

The test method is assessed by comparing its results to those obtained by means of another already validated test method, which has been developed for the same purposes. If this is not possible, the performance characteristics of the method may be assessed through interlaboratory comparisons. The method is "valid" if the results obtained by the different laboratories fall within the expected uncertainty limit. Deviations beyond such limits may indicate e.g. a lack of control of the influencing parameters. The causes of this behaviour should be clarified and the method is to be redefined accordingly. The interlaboratory comparison does not always provide a comprehensive validation of the representativeness of the method, which may be accurate and stable, though physically "wrong".

The acceptance procedure for new or modified test methods is either (i) determined internally in the laboratory (ii) agreed upon between the customer and the laboratory or (iii) accepted by the authorities and/or accreditation bodies. A higher degree of reliance is needed when safety, health and large economic values are involved. Calibration has been emphasized as an important element in the method validation process, but it is not necessarily the most dominating factor. The understanding of the testing method with its systematic and random errors is crucial. A scientific approach to analyze sources of error as well as the competence of the personnel doing that job is of great importance.

The laboratory should always describe the way the validation of test methods is done and this description should be a part of the quality system/manual when appropriate.

As simplified validation procedures ("fast" validation methods) must be used in many cases, the capability to use professional judgement in assessing whether the validation is comprehensive enough becomes pronounced. However, even when talking about simplified or fast validation procedures, the validation must be done with such a depth that the method is fit for the intended use and acceptable to the customer and/or authorities. It is clear that the definition of the use and scope of the method and assumption of uncertainty should not be misleading and too optimistic.

When the use of new test methods becomes more extensive, work describing the effect of changes in test parameters can be initiated in order to show the robustness of the method. Prenormative research should also be initiated.

The need for new or improved test methods arises when we lack methods or the existing ones are not complete, good or efficient enough. There is no need for the laboratory community to develop new methods if existing ones can be considered adequate.

Annex 1

Definitions

Repeatability (of results of measurement)

Closeness of the agreement between the results of successive measurements of the same measurand carried out under the same conditions of measurement. (VIM)

Notes

1. These conditions are called repeatability conditions
2. Repeatability conditions include:
   - the same measurement procedure
   - the same observer
   - the same measuring instrument, used under the same conditions
   - the same location
   - repetition over a short period of time
3. Repeatability may be expressed quantitatively in terms of the dispersion characteristics of the results.

Reproducibility (of results of measurements)

Closeness of the agreement between the results of measurements of the same measurand carried out under changed conditions of measurement. (VIM)

Notes

1. A valid statement of reproducibility requires specification of the conditions changes.
2. The changes conditions may include
   - principle of measurement
   - method of measurement
   - observer
   - measuring instrument
   - reference standard
   - location
   - conditions of use
   - time.
3. Reproducibility may be expressed quantitatively in terms of the dispersion characteristics of the results.
4. Results are here usually understood to be corrected results.

Uncertainty (of measurement)

Parameter, associated with the result of a measurement, that characterizes the dispersion of the values that could reasonably be attributed to the measurand. (BIPM/IEC/IFCC/ISO/IUPAC/IUPAP/OIML).

Notes

1. The parameter may be, for example, a standard deviation (or a given multiple of it), or the width of a confidence interval.
2. Uncertainty of measurement comprises, in general, many components. Some of these components may be evaluated from the statistical distribution of the results of series of measurements and can be characterized by experimental standard deviations. The other components, which can also be characterized by standard deviations, are evaluated from assumed probability distributions based on experience or other information.
3. It is understood that the result of the measurement is the best estimate of the value of the measurand, and that all components of uncertainty, including those arising from systematic effects, such as components associated with corrections and reference standards, contribute to the dispersion.

Validation

Confirmation by examination and provision of objective evidence that the particular requirements for a specific intended use are fulfilled (ISO 8402).

Notes

1. In design and development, validation concerns the process of examining a product to determine conformity with user needs.

2. Validation is normally performed on the final product under defined operating conditions. It may be necessary in earlier stages.
3. The term "validated" is used to designate the corresponding status.
4. Multiple validations may be carried out if there are different intended uses.

Verification

Confirmation by examination and provision of objective evidence that specified requirements have been fulfilled (ISO 8402).

Notes

1. In design and development, verification concerns the process of examining the result of a given activity to determine conformity with the stated requirements for that activity.
2. The term "verified" is used to designate the corresponding status.

Annex 2

References and background material

ISO 5725 (1986): Precision of test methods - Determination of repeatability and reproducibility for a standard test method by inter-laboratory tests.
ISO 8402 (1994): Quality management and quality assurance - vocabulary.
BIPM/IEC/IFCC/ISO/IUPAC/IUPAP/OIML (1995): Guide to the expression of uncertainty in measurement.
EAL-G 23 (1996): Expression on uncertainty in quantitative testing.
EAL-G 12 (1995): Traceability of measuring and test equipment to national standards.
Forstén Jarl (1991): A view on the assessment of the technical competence of testing laboratories. Nordtest Technical Report 149, 46p.
ILAC (1996): Guide an assessment and reporting of compliance with specification, based on measurements and tests in a laboratory.
VIM (1993): International vocabulary of basic and general terms in metrology.

Contact person:
M. Holmgren
EUROLAB Secretariat, SP Box 857,
S-50115 Borås, Sweden

Relevant accreditation policies and concepts in EU-, EFTA-, NAFTA- and other regions

Accred Qual Assur (1999) 4:45–49

Alison Gillespie
Sue Upton

Marketing Valid Analytical Measurement

Abstract The Valid Analytical Measurement (VAM) programme was set up by the Department of Trade and Industry as part of its support for the UK National Measurement System. This paper gives an overview of the VAM programme together with a description of the principles on which valid analytical measurement should be based. This is followed by a description of the work that has been carried out to market the results of the VAM programme to the analytical community.

Key words Valid analytical measurements · Marketing technology · Technology transfer

Introduction

Analytical measurement has a vital role in ensuring the quality of goods and commodities and in supporting Government in areas such as revenue collection, health and safety, environmental protection, agriculture and law enforcement. Clearly the data from analytical measurements need to be fit for their intended purpose. However, there is evidence to suggest that frequently there is lack of agreement on the validity of the data. One estimate puts the costs to British industry of unreliable data at at least ≤1 billion per year.

Following a review of its support for the National Measurement System (NMS) [1] in 1989, the Department of Trade and Industry (DTI) set up the Valid Analytical Measurement (VAM) programme, recognising for the first time the need to include analytical measurements within the NMS. Previously it had been concerned mainly with physical measurements.

The objectives of the NMS are: to enable individuals and organisations in the United Kingdom to make measurements competently and accurately and to demonstrate the validity of such measurements; to co-ordinate the UK's measurement system with measurement systems of other countries.

In the physical area the validity of measurements is established by means of an unbroken chain of comparisons from the working standards in day-to-day use through reference standards to the national and international primary standards, using recognised measurement techniques and methods. In this way the measurements are "traceable" back to the primary standards.

Unfortunately this infrastructure of working standards, reference standards and primary standards does not exist for analytical measurements. Although the Comité International des Poids et Mesures (CIPM) have started work on the development of such an infrastructure, it will take several years, if not decades, to develop. Thus the VAM programme is designed to meet the objectives of the NMS, i.e. to enable organisations to make measurements that are fit for their intended purpose, by:

the use of *validated methods* which are calibrated using *reference materials*, carried out with proper *quality assurance (QA) and quality control (QC) procedures*, including participation in *proficiency testing schemes*. With the laboratory methods and procedures being subjected to third-party *accreditation*.

The aim of the VAM programme is to encourage the use of the above procedures to improve the quality of the analytical measurements in the United Kingdom. The work centres around three main areas of activity: defining and disseminating best analytical practice which will enable laboratories to deliver reliable results every time; developing the tools which enable laboratories to implement best analytical practice; working with analysts in other countries to ensure the comparability of analytical measurements across international boundaries.

Best practice through the VAM principles [2]

Best analytical practice has been defined by six VAM principles (see Table 1). Valid measurements and agreement between laboratories can be achieved by adhering to these basic principles. The VAM principles describe an approach which is consistent with the best of modern quality systems, presented in a way which is directly meaningful to those making analytical measurements. The principles are fundamental to VAM and are designed to control all factors that might affect the reliability of analytical results. Unlike formal quality management systems there is no accreditation process and as such the philosophy of VAM is similar to that of a total quality management (TQM) system for analytical laboratories, which can be applied whatever the size of organisation or nature of work. The objective of the principles was to raise awareness of the importance of obtaining reliable "fit for purpose" measurements.

Addressing the aims of VAM

The work carried out under the VAM programme covers three broad technical themes – chemical measurement, physical measurement and biologically based analytical measurement. LGC is the lead contractor and works together with the National Physical Laboratory (NPL) and Atomic Energy Authority Technology (AEAT) in delivering the VAM programme.

Table 1 The six VAM principles of best analytical practice

1. Analytical measurements should be made to satisfy an agreed requirement.
2. Analytical measurements should be made using methods and equipment which have been tested to ensure they are fit for purpose.
3. Staff making analytical measurements should be both qualified and competent to undertake the task.
4. There should be a regular independent assessment of the technical performance of a laboratory.
5. Analytical measurements made in one location should be consistent with those elsewhere.
6. Organisations making analytical measurements should have well-defined quality control and quality assurance procedures.

Currently there are several technical projects being carried out at the Laboratory of the Government Chemist (LGC) in the areas of education and training, reference materials, quality systems and management, confidence and validation, out-of-laboratory measurement, DNA technology research and high accuracy chemical analysis. A number of these areas are covered in other articles in this issue of the journal. Two of the more recent areas of work that are creating a lot of interest are DNA technology research and out-of-laboratory measurement.

DNA technology, in particular, is having a revolutionary effect on a host of industrial and regulatory sectors. This area is rapidly developing and offers tremendous advantages and benefits to industry, but there is an urgent need for parallel validation of the analytical techniques employed in DNA-based measurements and development of tools to enhance validity such as suitable reference standards. Analytical molecular biology has typically been developed, and is most often employed, in academic and medical research environments where there is little need to consider the more routine applicability, reliability and reproducibility of the methods. Evaluation of these factors and further validation of the methods is therefore necessary, particularly when such techniques are applied to the analysis of "real" samples.

Studies undertaken by LGC during the VAM 1994–1997 programme have demonstrated that a large number of generic problems need to be addressed before the potential power of nucleic-acid-based methods can be reliably applied. The main outputs of the DNA technology validation project in the new programme will be validated novel measurement systems which utilise DNA technological procedures, quality protocols, and reference materials.

Out-of-laboratory measurements are undertaken across a broad range of industrial and analytical sectors for a variety of reasons: in clinical and medical diagnostics; for the control of chemical and petrochemical production processes; and to monitor emissions and discharges to the environment. The validity of data derived from such measurements is clearly of vital importance, for example to demonstrate compliance with environmental legislation. However, it is particularly difficult to obtain valid and reliable measurements outside the laboratory. The inability to control the environment in which the measurements are made and the use of untrained operators both have potential to impact significantly on the reliability of data. The situation is made worse because of the lack of adequate QA and QC procedures, the shortage of reference materials and calibration standards, and the lack of adequate specifications and performance data for the procedures and equipment employed.

The current state of practice also varies considerably across sectors. Under the VAM programme, the aims are to increase user confidence in the validity of data derived from portable measuring equipment and to improve the reliability and comparability of measurements made using portable equipment. The possibility of establishing formal QA and QC procedures and proficiency testing for out-of-laboratory measurements will also be explored.

Marketing VAM

A major part of the VAM programme has been the marketing and technology transfer of the work carried out to the market place itself. Marketing VAM poses particular challenges to the marketer in that it is a philosophy rather than a tangible product or a service. The intangibility of VAM also makes it a difficult concept for organisations to understand and sign up to. Promotion, and ultimately uptake, of VAM would be more effective, if ways could be identified to make it more tangible.

Marketing strategy

The key starting point for the effective marketing of VAM was to agree on a product definition, proposition and positioning for VAM. The final definitions arrived at are outlined in Table 2 and describe the VAM product being marketed, the benefits of adopting VAM (namely reduced costs and risks to an organisation) and the positioning of VAM relative to other quality systems (self-regulated implementation). In addition to defining the VAM product it was also recognised that there was a need to secure adoption of the principles, rather than just awareness of them and to identify the barriers to adoption of VAM. The target market also needed to be tracked through the various stages from adoption to implementation.

Table 2 VAM product definitions

Product description
VAM sets out six principles of good analytical practice, backed up by technical support and management guidance.

Product positioning
Self-regulated implementation of best practice for analytical science to produce reliable consistent and comparable results for measurable commercial gain.

Selling proposition
Adoption of the principles of VAM will reduce the costs and risks associated with unreliable measurements.

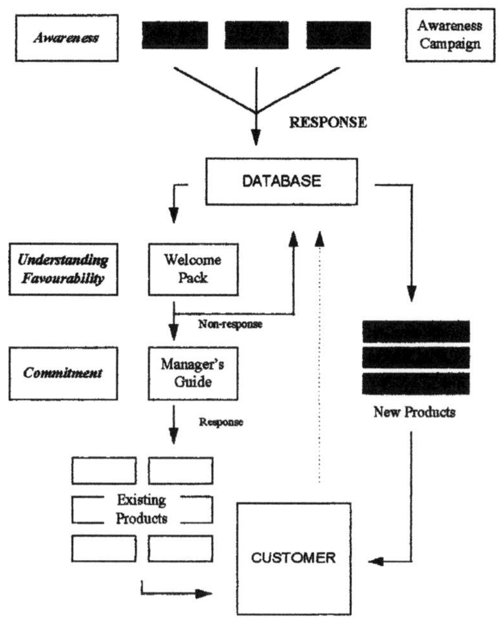

Fig. 1 The VAM promotional model

A promotional model to be used as a framework for the marketing of VAM was developed to address the issues outlined above (see Fig. 1). Development of the model also took place alongside a quantitative telemarketing survey conducted amongst identified decision makers in industry. The objectives of the survey were to: identify potential barriers to adoption of VAM; identify the communication objectives and key messages; benchmark the overall awareness of VAM.

The model indicated the various stages that a VAM "prospect" needed to move through towards adoption of VAM and attempted to define the communication objectives for each stage. For example, the initial objective is to create awareness of VAM amongst the target audience. Following on from this the task is to create understanding and an attitude of favourability towards VAM. The final step towards adoption of VAM is an expression of commitment to implementation. In the model, adoption of VAM is measured by uptake of VAM products, for example books, videos and certified reference materials. In effect the VAM products become the tools to aid implementation of the six VAM principles. The right-hand side of the model indicates that the new VAM clients, or implementers, also form a community that will create a requirement for future VAM products and services.

The target audience for the new VAM marketing strategy was defined as senior decision-makers in industry, that is those persons towards the top of an organisation with influence to change attitudes, work practices and realise the commercial benefits of adoption. The need to identify "early adopters" or "champions" for VAM was identified.

In order to implement the strategy a range of new materials were produced, namely:

1. VAM direct mail piece, used to raise awareness of VAM and elicit request for further information.
2. VAM welcome pack "Better Measurements Mean Better Business", which sets out the business benefits of adopting VAM to senior decision makers. This was produced to improve understanding and favourability towards VAM following an expression of interest.
3. "Managers Guide to VAM". A more detailed guide on how to implement VAM in the analytical laboratory. The guide includes a self-assessment exercise, used to highlight priority areas for immediate attention and details sources of further help and advice.
4. New VAM database used to track prospects from initial point of contact through to adoption or rejection of VAM.

Implementation of the VAM promotional model

Implementation of the model began with a campaign which aimed to generate awareness amongst senior decision-makers of the costs and risks of unreliable measurements. A direct mail campaign was commenced aimed at senior decision-makers in industry. In the first phase the conversion rate from direct mailshot to request for the VAM welcome pack was slightly higher than might be expected (7%) and the conversion rate from welcome pack to the Managers Guide to VAM was encouraging (35%), suggesting the message seemed to be striking a chord. A follow-up telemarketing campaign was also instigated which resulted in an increase in response rate as a result of the more personalised nature of the contact. However, it was clear that many difficulties arose because of the poor quality of data contained on most databases. In this respect the effectiveness of direct mail as a technique for generating awareness and response proved difficult to evaluate. There is a requirement to obtain either a better-quality database and repeat the direct mail exercise, or to identify an alternative mechanism for communicating with senior decision-makers.

The model will only work if there is a convincing business case for the adoption of VAM. In practice so far this has proved quite difficult to demonstrate. High profile cases in the public domain have high impact but are often perceived to be politically sensitive. In the search for exemplar organisations there is a general unwillingness for companies to admit that bad practice was ever present in their organisation, no matter how much they have since improved. However, this whole area is a key to the recruitment of both producers and users to VAM and needs to be addressed.

The model identifies a mechanism by which VAM is adopted by producers of data. However, it falls short of explaining how communication with the bench analyst will take place and what should be the nature of this type of communication. In practice it is actually this level that the VAM contractors are most experienced at communicating since generally the content is technically, as opposed to commercially, biased.

Measurement of the adoption of VAM is not easy. Whilst uptake of VAM products and CRMs is one measure, in practice this information is difficult to pull together for an individual, especially when products can be purchased from a number of suppliers. An assumption has to be made that once a product is purchased it will also be used.

Future plans

By definition, it is proposed that the future plans for promoting VAM to producers of analytical data should concentrate on "the individual or department with responsibility for the production and overall quality of an analytical measurements". In some cases these individuals may be senior decision makers but in the majority of cases these are more likely to be bench analysts, quality managers and laboratory team leaders. This suggests that an appropriate base for segmentation of the "producer" audience for VAM is by job function, with the target segments ranging from those individuals who actually make the measurements (producer analysts) to those who have overall responsibility for their quality (senior decision-makers), but who may not actually carry out any analysis themselves.

Recognising that the producers of analytical data can be segmented as such enables the benefits of VAM to be positioned more appropriately for each target segment. Whilst the overall benefit of VAM to the producers of analytical data can be described as reduction in the costs and risks associated with unreliable measurements, this "business benefits" message is most likely to be of interest to the senior decision makers. The producer analyst is much more likely to be interested in the technical aspects of the VAM programme, i.e. their job is concerned with "how" to implement VAM as opposed to "why". An interest in the "how" part of the adoption is likely to arise when: there is pressure to improve quality from within the organisation itself, i.e. via senior decision-makers convinced of the business benefits (self-regulation argument); there is pressure to improve quality from internal or external customers (user demand); the individual concerned is enlightened enough to see the benefits of adoption in their work without any external pressure being applied.

Whilst senior decision-makers who are producers of analytical data can still be encouraged to adopt VAM through the route suggested in the VAM promotional model, i.e. via the VAM welcome pack and manager's guide, the model needs further adaptation in order to explain how the producer analysts can be addressed.

Promotion to users of analytical data

Over the last year or so, there has been increasing recognition that another important target market for VAM, as well as "producers" of analytical data, is the "users" of analytical data. Demand from these so-called users for quality in analytical measurements is believed to be an important driver towards improving the overall quality of analytical measurements made in the United Kingdom.

Essentially, there are two mechanisms by which the quality of analytical measurements made in the United Kingdom can be improved: by the producers improving the quality of the analytical measurements they supply; by the users of analytical measurements demanding an improvement in the quality of analytical measurements they procure.

Past thinking was based on the premise that it was preferable to equip the supplier of data with the tools required to satisfy demand, before stimulating demand itself, and that the inherent value of quality would be a concept readily appreciated by such parties. Thus VAM has been promoted to producers of analytical data as self-implemented best practice, which will bring about measurable commercial gain to an organisation through a reduction in the costs and risks associated with unreliable measurements. However, without an external driver for adoption it is likely that some laboratories will simply choose to ignore VAM, taking a calculated risk or, in some cases, simply dismissing the issue altogether. There is a cost associated with implementing quality systems (both up-front and ongoing) and if those costs are to be justified the producers of data need to be able to realise the benefits as well. Users of analytical data also need to understand the implications when purchasing data in terms of an additional price they might be expected to pay.

Fundamental to the marketing of VAM to users is the need to describe what exactly is meant by a user of analytical data. There are a range of individuals or organisations who might commission analytical measurements directly from these providers. Organisations, such as banks and insurance companies, also rely on the results of analytical measurements, for example in risk assessment, without actually dealing with the producer of the measurements themselves, i.e. they operate through a third party. In addition there are organisations and individuals (e.g. consumers) who, perhaps without even knowing, depend on the results of analytical measurements.

It seems logical, therefore, to identify producers and users of analytical data according to their position in a supply chain of analytical data, which begins at the point where the measurement is made and leads up to the point where it is actually put to use. It is also important to recognise that:

1. a supply chain can exist within an organisation or between two or more organisations and individuals;
2. at different positions in the supply chain there will be different decisions

taken and these may vary in the level of risk;
3. it is possible for an individual to fulfil two roles in the supply chain, for example as a user commissioning data and subsequently as a provider of the data (possibly with some value added interpretation) to the next individual or organisation in the chain;
4. the more remote an individual or organisation is from the producer, the less likely they are to be aware of the issues concerning quality in analytical measurement. In effect this means they are more likely to take analysis for granted or simply not even consider it at all.

A suggested definition for producers and users of analytical data is as follows:

A *producer* of analytical measurements is "the individual or department with responsibility for the production and overall quality of an analytical measurement".

A *user* of analytical data is "an individual or group who makes a decision based on the results of an analytical measurement, be they internal or external to the producer organisation".

These definitions imply that the "senior decision-makers" targeted previously in the promotional model for producers could in fact be both producers and users of analytical data. It is clear that, to market VAM effectively, a clear understanding of the benefits of VAM is required from the distinct viewpoint of users and producers. In the case of senior decision-makers they will be in a position to take advantage of some or all of these dependent on whether they are adopting a user or producer role.

Benefits of VAM to users and producers

In the VAM welcome pack and the manager's guide the benefits of VAM are presented as a reduction in costs (both direct and hidden) and risks. These are defined as:
1. The direct costs of unreliable measurements are those costs incurred if an analysis has to be repeated because the original measurements are known, or suspected to be unreliable.
2. The hidden costs of unreliable measurements are the costs which are hidden internally in the organisation, because the true impact of unreliable measurements is not recognised.
3. The risks of unreliable measurements are the risks associated with the release of unreliable measurements to a customer.

In recognising the distinction between a user and producer of analytical measurements and attempting to clarify the benefits of VAM to both parties, it is clear that the above definitions, whilst providing a useful framework for promoting VAM to producer organisations, do not go far enough. The reality of the situation is that the direct costs of unreliable measurements are only relevant to the producers of analytical measurements (unless they are passed on as an increased cost to the user, in which case the user needs to be aware of what they are or are not paying for). For users of analytical data the main concern is one of risk from poor decision-making.

Current situation

Promotion to users of analytical data is a complex and difficult task. Users are less likely to be aware of or educated in the issues relevant to the technically complex matters which affect quality in analytical measurement. They are also more difficult to identify and reach.

There is a body of anecdotal evidence that suggests awareness of VAM, and best practice in analytical measurement amongst users of analytical data is low. Concern has been expressed regarding the way in which potential customers went about choosing a contract laboratory. Factors such as quality of data were rarely considered, with the exception of a request in some instances for the laboratory to be NAMAS (National Accreditation of Measurement and Sampling) accredited. Cost appears to be the main factor considered, with customers often selecting cheaper, less quality conscious laboratories, without consideration of the associated risks.

It is important to also recognise that whilst the risks associated with the use of unreliable measurements are undoubtedly real, there is a counter argument which says that most users have survived or managed that risk to date. This may be because the true impact has not been felt yet as, for example, in the case of contaminated land analysis where there is a long time delay between the analysis being carried out and the consequences of poor quality analysis being felt. Also, analysis of samples may be repeated several times as a means of reducing risk. If the true factors impacting risk could be calculated then a more appropriate, "fit for purpose", measurement could be carried out (possibly even at a reduced cost).

The primary objective of the promotion to users of analytical data is to improve the quality of analytical measurements made in the United Kingdom by stimulating demand for VAM from the users of analytical data. In developing a strategy to meet this objective it is recognised that VAM needs to be made more tangible and visible, and clear business benefits need to be identified and communicated persuasively to the relevant business audience.

References

1. Measuring up to the competition (1989) Her Majesty's Stationery Office, London. Cm 728
2. VAM Bulletin, 13:4–5 (Autumn 1995)

A. Gillespie (✉) · S. Upton
The Laboratory of the Government Chemist,
Queens Road,
Teddington, TW11 OLY, UK
e-mail: amg@lgc.co.uk
Tel.: +44-181-9437000
Fax: +44-181-943 2767

Accred Qual Assur (2002) 7:294–298
DOI 10.1007/s00769-002-0484-9

Rouvim Kadis

Analytical procedure in terms of measurement (quality) assurance

Abstract In the ordinary sense the term "analytical procedure" means a description of what has to be done while performing an analysis without reference to quality of the measurement. A more sound definition of "analytical procedure" can be given in terms of measurement (quality) assurance, in which a specified procedure to be followed is explicitly associated with an established accuracy of the results produced. The logic and consequences of such an approach are discussed, with background definitions and terminology as a starting point. Close attention is paid to the concept of measurement uncertainty as providing a single-number index of accuracy inherent in the procedure. The appropriateness of the uncertainty-based approach to analytical measurement is stressed in view of specific inaccuracy sources such as sampling and matrix effects. And methods for their evaluation are outlined. The question of a clear criterion for analytical procedure validation is also addressed from the standpoint of the quality requirement which measurement results need to meet as an end-product.

Keywords Accuracy · Analytical procedure · Measurement assurance · Measurement uncertainty · Quality assurance · Validation

Introduction

There are different ways in which quality assurance concepts play a role in analytical chemistry. Most of them such as stipulating requirements for the competence and acceptance of analytical laboratories, writing quality manuals, performing systems audits, etc. can be viewed as something foreign to common analytical thinking forced upon analysts by external authorities. Perhaps another possible way is to try to integrate quality assurance aspects into common analytical concepts and (re)define them in such a way as to explicitly include the quality matters required. This may facilitate an effective quality assurance strategy in analytical chemistry.

In ordinary usage the term "analytical procedure" hardly needs a referential definition and may be for this reason there are few official definitions of the term. The only definition quoted in the references [1] is rather diffuse:

"The analytical procedure refers to the way of performing the analysis. It should describe in detail the steps necessary to perform each analytical test. This may include but is not limited to: the sample, the reference standard and the reagents preparations, use of the apparatus, generation of the calibration curve, use of the formulae for the calculation, etc." [2].

In brief, this simply means a description of all that should be done in order to perform the analysis.

Leaving aside some prolixity in the definition above, the main thing that is lacking is the goal requirement needed in considering quality matters. As it is shown in this paper, a sound definition of an analytical procedure can be given in terms of measurement (quality) assurance. The case in point is not simply "the way of performing the analysis" but that which ensures obtaining the results of a specified quality. What this eventually means is a prescribed procedure to follow in producing results with a known uncertainty.

If we have indeed recognized chemical analysis to be measurement, though possessing its own peculiarities, we can apply the principles and techniques of quality assurance developed in measurement to analytical work. These principles and techniques constitute the field of measurement assurance [3], a system affording a confidence that all the measurements produced in a measurement process maintained in statistical control are good enough for their intended use. "Good enough" implies here nothing more than having an allowable uncertainty. Although measurement assurance was originally developed for instrument calibration, i.e. with emphasis on measurement traceability, it is reasonable to treat it more generally. One can say that a fixed measurement procedure is a means of assigning an uncertainty to a single measurement, and this is the essence of measurement assurance. This also reveals the role a prescribed (analytical) procedure plays in routine analytical measurement. We will focus on different aspects involved in the concept of an analytical procedure defined in terms of measurement assurance such as terminology, content, evaluation, and validation.

Starting point

Chemical analysis generally consists of several operational stages beginning with taking a sample representative of the whole mass of the material to be analysed and ending with calculation and reporting of results. In this sequence the measurement proper usually makes a relatively small contribution to the overall variability involved in the entire chemical measurement process (CMP) [4], the largest portion of which being concerned with "non-measurement" operations such as isolation, separation, and so on. Because of this, everything in the chain that affects the chemical measurement result must be predetermined as far as practically possible: the experimental operations, the apparatus and equipment, the materials and reagents, the calibration and data handling. Thus, a "*complete analytical procedure*, which is specified in every detail, by fixed working directions (order of analysis) and which is used for a particular analytical task" [5] – a concept presented by Kaiser and Specker as far back as in the 1950s [6] – becomes a point of critical importance in obtaining meaningful and reproducible results. We use the term "analytical procedure" or merely "procedure" for short, in the sense outlined above.

"Method", "procedure", or "protocol"

The importance of the correct usage of relevant terms, in particular, the term "procedure" rather than "method" is noteworthy. The terms actually correspond to different levels in the hierarchy of analytical methodology [7] expressed as a sequence from the general to the specific:

technique ⇒ method ⇒ procedure ⇒ protocol

Indeed, the procedure level provides the specific directions necessary to utilize a method, which is in line with the definition of *measurement procedure* given in the International Vocabulary of Basic and General Terms in Metrology (VIM): "set of operations, described specifically, used in the performance of particular measurements according to a given method" [8].

This nomenclature is however not always adhered to. In many cases, i.e. scientific publications, codes of practice, or official directives, an analytical procedure is virtually implied when an analytical method is spoken about. Commonly used expressions such as "validation of analytical methods" or "performance characteristics of analytical methods" are typical examples of incorrect usage. Such confusion appears even in the definition suggested by Wilson in 1970 for the term "analytical method" [9]. As he then put it, "an analytical method is to be regarded as the set of written instructions completely defining the procedure to be adopted by the analyst in order to obtain the required analytical result". It is actually difficult to make a distinction between the two notions when one of them is defined in terms of the other.

On the other hand, there is normally no reason to differentiate the two most specific levels in the hierarchy above, carrying the term "procedure" over to the designated "protocol". The latter was defined [7] as "a set of definitive directions that must be followed, without exceptions, if the analytical results are to be accepted for a given purpose". So, written directions have to be faithfully followed in both cases. In many instances the term "procedure" actually signifies a document in which the procedure is recorded – this is specifically noted in VIM in respect to "measurement procedure". Besides, the term "standard operating procedure" (SOP), especially applied to a procedure intended for repetitive use, is popular in quality assurance terminology.

A clear distinction needs to be drawn between analytical procedure as a generalized concept and its particular realization, i.e. an individual version of the procedure arising in specific circumstances. In practice, an analytical procedure exists as a variety of realizations, differing in terms of specimens, equipment, reagents, environmental conditions, and even the analyst's own routine. Not distinguishing between these concepts can lead to a misinterpretation embodied, for instance, in the viewpoint that with a detailed specification the procedure will change "each time the analyst, the laboratory, the reagents or the apparatus changed" [9]. What will actually change is realizations of the procedure, only provided that all specified variables remain within the specification.

Also one cannot but note that the hierarchy of methodology above concerns, in fact, a level of specificity rather than the extent to which the entire CMP may be covered. Although sampling is the first (and even the most critical) step of the process, it is often treated as a separate issue when addressing analytical methodology. A "complete analytical procedure" may or may not include sampling, depending on the particular analytical problem to be solved and the scope of the procedure.

An analytical procedure yields the results of established accuracy

In line with Doerffel's statement which refers to analytical science as "a discipline between chemistry and metrology" [10], one may define analytical service – as a sort of analytical industry, that is practical activities directed to meeting customer needs – as based upon concepts of chemistry, metrology, and industrial quality control. The intention of any analytical methodology in service is to produce data of appropriate quality, i.e. those that are fit for their intended purpose. The question to answer is what kind of criteria should be addressed in characterizing fitness-for-purpose.

From the viewpoint of objective of measurement, which is to estimate the true value of the quantity measured, and its applicability for decision-making, closeness of the result to the true value, no matter how it is expressed, should be such a criterion. If a measurement serves any practical need, it is to meet an adequate level of accuracy. It is compliance with an accuracy requirement that fundamentally defines the suitability of measurement results for a specific use, and hence corresponding demands are to be made on a measurement process that produces the results. Next it is assumed that the process is generated by the application of a measurement procedure and thus, the accuracy requirements should be finally referred to in the procedure itself. (The "requirements sequence" first implies substantiation of the demands on accuracy in a particular application area, the problem that needs special consideration in chemical analysis [11].)

Following this basic pattern, it is reasonable to re-define Kaiser's "complete analytical procedure", so that the fitness-for-purpose criterion is explicitly allowed for. There must be an accuracy constraint built in the definition so as to give a determining aspect of the notion. It is probably unknown to most analytical chemists worldwide that such a definition has long since been adopted in analytical terminology in Russia. This was formulated in 1975 by the Scientific Council on Analytical Chemistry of the Russian Academy of Sciences. As defined by the latter an *analytical procedure* is the: " a detailed description of all the conditions and operations that ensure established characteristics of trueness and precision" [12]. This wording which goes beyond the scope of analytical chemistry specifically differs from the VIM definition of measurement procedure quoted above by including an accuracy requirement as a goal function. It clearly points out that adhering to such a fixed procedure ensures (at least conceptually) that the results obtained are of a guaranteed degree of accuracy.

Two basic statements underlie the definition above. First, an analytical procedure when performed as prescribed, with the chemical measurement process operating in a state of control, has an inherent accuracy to be evaluated. Second, a measure of the accuracy can be transferred to the results produced, providing a degree of their confidence. In essence, the measure of accuracy typical of a given procedure is being assigned to future results generated by the application of the procedure under specified conditions. The justification for both the propositions was given by Youden in his work on analytical performance [13, 14] where methods for determining accuracy in laboratories were discussed in detail.

As a prerequisite for practical implementation of the analytical procedure concept it is assumed that the chemical measurement process remains in a state of statistical control, being operated within the specifications. To furnish evidence for this and to avoid the reporting of invalid data the analytical system needs to be tested for continuing performance. A number of control tests may be used with this aim, for instance, testing the difference between parallel determinations when they are prescribed to be carried out, duplicating complete analysis of a current test material, and analysis of a reference material. An important point is that the control tests are to be performed in such a manner and in such a proportion as a given measurement requires, and are an integral part of the whole analytical procedure. The control tests may be more specific in this case and relate to critical points of the measurement process. As examples, calibration stability control with even one reference sample or interference control by spiking provide useful means of expeditious control in an analytical procedure.

A principal point in this scheme is that accuracy characteristics should be estimated before an analytical procedure is regularly used and should be the characteristics of any future result obtained by application of the procedure under specified conditions. Measurements of this type are most commonly performed (by technicians, not measurement scientists) in control engineering and are sometimes called "technical measurements". It is such measurements that are usually referred to as routine in chemical analysis. In fact, the problems of evaluation of routine analyses faced by chemists are treated more generally in the "technical measurements" theory [15].

Uncertainty as an index of accuracy of an analytical procedure

It is generally accepted that accuracy as a qualitative descriptor can be quantified only if described in terms of precision and trueness corresponding to random and systematic errors, respectively. Accordingly, the two measures of accuracy, the estimated standard deviation and the (bounds for) bias, taken separately, have to be generally evaluated and reported [16]. As the traditional theory of measurement errors holds, the two figures cannot be rigorously combined in any way to give an overall index of (in)accuracy. Notice that accuracy, as such, ("closeness of the agreement between the result of a measurement and a true value of the measurand" [8]) by no means involves any measurement error categorization.

On the other hand, it has long been recognized that the confidence to be placed in a measurement result is conveniently expressed by its uncertainty that was thought, from the outset, to mean an estimate of the likely limits to the error of measurement. So, uncertainty has traditionally been treated as "the range of values about the final result within which the true value of the measured quantity is believed to lie" [17]. However, there was no agreement on the best method for assessing uncertainty. Consistent with the traditional subdivision, the "random uncertainty" and the "systematic uncertainty" each arising from corresponding sources should be kept separate in the evaluation of a measurement, and the question of how to combine them was an issue of debate for decades.

Now a unified and widely applicable approach to the uncertainty statement set out in ISO Guide (GUM) [18] is being accepted in many fields of measurement, particularly in analytical measurements due to the helpful adaptation in the EURACHEM Guide [19]. Some peculiarities of the new approach can be intimated, specifically, the abandonment of the previous distinction between random and systematic uncertainties, treating all of them as standard-deviation-like quantities (after the corrections for known systematic effects have been made), and their possible estimation by other than statistical means. Fundamental, however, is that any numerical measurement is not thought of in isolation, but in relation to the process which generates the measurements. All the factors operative in the process being defined, they virtually determine the relevant uncertainty sources, so making practicable their quantification to finally derive the value of total uncertainty. One can say that the measurement uncertainty methodology fits neatly the starting idea of a procedure specified in every detail, since the procedure itself defines the context which the uncertainty statement refers to.

This is true of the component-by-component ("bottom-up") method for evaluating uncertainty that is directly in line with GUM. Also this is true for the "top-down" approach [20] that provides a valuable alternative when poorly understood steps are involved in the CMP and a full mathematical model is lacking. An important point is that the top-down methodology implies a reconciliation of information available with the required one that is based on a detailed analysis of the factors which affect the result. For both approaches to work advantageously a clear specification of the analytical procedure is evidently a necessary condition.

The break with the traditional subdivision of measurement errors has a crucial impact on the way accuracy may be quantified and expressed. In 1961, Youden wrote [14]: "There is no solution to the problem of devising a single number to represent the accuracy of a procedure". He was indeed right in the sense that a strict probability statement cannot be made about a combination of random and systematic errors. Today, thanks to the present uncer-

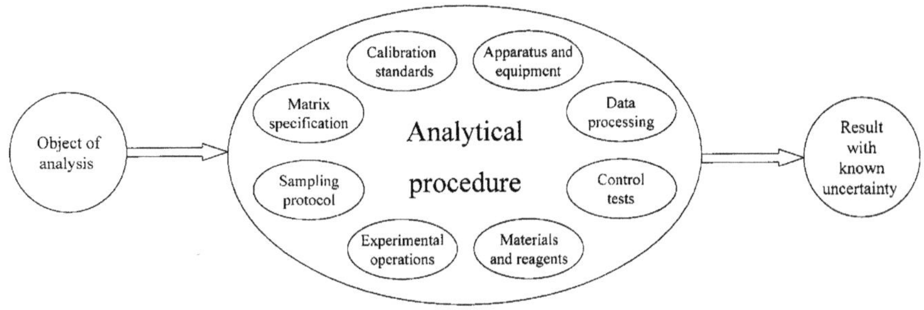

Fig. 1 Typical "constituents" to be specified within analytical procedure, which ensures obtaining the results with a known uncertainty

tainty concept, we maintain the other opinion that such a solution does exist. It is measurement uncertainty that can be regarded as a single-number index of accuracy inherent in the procedure. In doing so we must not be confused by the fact that the operational definition of measurement uncertainty that GUM presents does not use the unknown "true value" of the measured quantity following pragmatic philosophy. The old definitions and, in particular, that cited above are equally valid and are now considered ideal.

Consequently, we can define an analytical procedure as leading to results with a known uncertainty, as in Fig. 1 in which typical "constituents" to be specified in an analytical procedure are shown.

Specific inaccuracy sources in an analytical procedure

What has been said in the previous section generally refers to specified measurement procedures used in many fields of measurement. There are, however, some special reasons, specific to chemical analysis, that make the uncertainty methodology particularly appealing in analytical measurements. This is because of specific inaccuracy sources in an analytical procedure which are difficult to be allowed for otherwise. Two such sources, sampling and matrix effects, will be mentioned here, with an outline of the methods for their evaluation.

Sampling

Where sampling forms part of the analytical procedure, all operations in producing the laboratory sample such as sampling proper, sample pre-treatment, carriage, and sub-sampling require examination in order to be taken into account as possible sources contributing to the total uncertainty.

It is generally accepted that a reliable estimate of this uncertainty can be obtained empirically rather than theoretically. Accordingly, an appropriate methodology has being developed [e.g. 21, 22] aimed at separating the sampling contribution from the total variability of the measurement results in a specially designed experiment. This is not, however, the only way of quantifying uncertainty in sampling. Explicit use of scientific judgement is now equally approved when experimental data are unavailable. An illustrative example from the EURACHEM Guide (Ref. 19, Example A4) clearly demonstrates the potential of mathematical modelling inhomogeneity as an alternative to the sampling assessment experiment.

It is significant that with the uncertainty methodology both the major analytical properties, "accuracy" and "representativeness" [23], which quality of analytical data relies on, can be quantified and properly taken into account to give a single index of accuracy. This index expresses consistency between the measurement results and the true value that refers to a bulk sample of the material rather than the test portion analysed.

Matrix effects

The problem of matrix mismatch is always attendant when one analyses an unknown sample "with the same matrix" using a fixed, previously determined, calibration function. Not uncommonly, an analytical procedure is developed to cover a range of sample matrices in such a way that an "overall" calibration function can be used. An error due to matrix mismatch is therefore inevitable if not necessary significant. Commonly regarded as systematic for a sample with a particular matrix, the error becomes random when a population of samples to which the procedure applies is considered; this in fact constitutes an inherent part of the total variability associated with the analytical procedure.

Meanwhile, these effects are in no way included in the usual measures of accuracy as they result from a "method-performance study" in accordance with the accepted protocols [24, 25]. The *accuracy experiment* defined by ISO 5725 (Ref. 24, Part 1, Section 4) does not presuppose any variable matrix-dependent contribution, being confined to *identical test items*. The underlying statistical model assumes that solely laboratory components of bias and their distribution must be considered.

It is notable that such kinds of error sources are fairly treated using the concept of measurement uncertainty which makes no difference between "random" and "systematic". When simulated samples with known analyte content can be prepared, the effect of the matrix is a matter of direct investigation in respect of its chemical composition as well as physical properties that influence the result and may be at different levels for analytical samples and a calibration standard. It has long since been suggested in examination of matrix effects [26, 27] that the influence of matrix factors be varied (at least) at two levels corresponding to their upper and lower limits in accordance with an appropriate experimental design. The results from such an experiment enable the main effects of the factors and also interaction effects to be estimated as coefficients in a polynomial regression model, with the variance of matrix-induced error found by statistical analysis. This variance is simply the (squared) standard uncertainty we seek for the matrix effects.

In many ways, this approach is similar to ruggedness testing aimed at the identification of operational (not matrix-related) conditions that are critical to the analytical performance.

"Method validation" in terms of measurement assurance

The presented concept of analytical procedure offers a clear perspective on the problem of "method validation" which is an issue of great concern in quality matters. Validation is generally taken to mean a process of demonstration that a methodology is suitable for its intended application. The question is how should suitability be assessed, based on customer needs?

It is commonly recommended [e.g. 2, 28–30] that a number of characteristics such as selectivity/specificity, limits of detection and quantitation, precision and bias, linearity and working ranges be considered as criteria for analytical performance and evaluated in the course of an validation study. In principle, they need to be compared to some standard; based on this, judgement is made as to whether the *procedure* under issue is capable of meeting the specified analytical requirements, that is to say, whether a "method is fit-for-purpose" [28].

However, from the perspective of end-users of analytical results, it is important that the *data* be only of the required quality and thus appropriate for their intended purpose. In other words, the matter of primary concern is quality of analytical results as an end-product. In this respect, a procedure will be deemed suitable when the data produced are fit-for-purpose.

It follows that the common criteria of validation should be made more specific in terms of measurement assurance. It is (the index of) accuracy that requires overriding consideration among the characteristics of analytical performance if quality of the results is primarily kept in mind. Other performance characteristics are desirable to ensure that a methodology is well-established and fully understood, but *validation* of an analytical procedure on those criteria seems impractical also in view of the lack of corresponding requirements as is commonly the case. (Strictly speaking, there is no validation unless a particular requirement has been set.)

We have every reason to consider the estimation of measurement uncertainty in an analytical procedure followed by the judgement of compliance with a target uncertainty value as a kind of validation. This is in full agreement with ISO 17025 that points to several ways of validation, among them "systematic assessment of the factors influencing the result" and "assessment of the uncertainty of the results..." [31]. In line with this is also a statistical modelling approach to the validation process that has recently been developed and exemplified as applied to in-house [32] and interlaboratory [33] validation studies.

A concrete example of such validation is worthy of notice. *Certification* (*attestation*) of analytical procedures used in regulated fields such as environmental control and safety is operative in the Russian state measurement assurance system as a process of establishing metrological properties and confirming their compliance with relevant requirements. (By metrological properties we mean herein the assigned measurement error characteristics, i.e. measurement uncertainty.) This is introduced by the Russian Federation state standard GOST R 8.563 [34] which also covers procedures for quantitative chemical analysis. This certification is, in fact, a legal metrology measure similar, to some extent, to pattern evaluation and approval of measuring instruments. Some scepticism concerning the efficiency of legal metrology practice in ensuring the quality of analytical measurements may be in order. Nevertheless, the conceptual (measurement assurance) basis of this approach to validation deserves attention beyond doubt.

Conclusions

This debate allows the following propositions to be made:

1. The term "analytical procedure" commonly used without reference to the quality of data is best defined in terms of measurement (quality) assurance to explicitly include quality matters. This means a specified procedure which ensures results with an established accuracy.
2. The measurement uncertainty methodology neatly fits the idea of a specified measurement procedure and furthermore provides a tool for covering specific inaccuracy sources peculiar to analytical measurement. Uncertainty can be regarded as a single-number index of accuracy of an analytical procedure.
3. When an analytical procedure is so defined, uncertainty becomes the performance parameter that needs overriding consideration over and above all the others assessed during validation studies. This kind of validation gives a direct answer to the question whether the data produced are of required quality and thus appropriate for their intended use.

References

1. Holcombe D (1999) Accred Qual Assur 4: 525–530
2. International Conference on Harmonization of Technical Requirements for Registration of Pharmaceuticals for Human Use (1994) Text on validation of analytical procedures. ICH Quality topic Q2A: Definitions and terminology (http://www.ifpma.org/ich5q.html)
3. Cameron JM (1976) J Qual Technol 8: 53–55
4. Currie LA (1978) Sources of error and the approach to accuracy in analytical chemistry. In: Kolthoff IM, Elving PE (eds) Treatise on analytical chemistry. Part I. Theory and practice, vol.1, 2nd edn. Wiley, New York, pp 95–242
5. Kaiser H (1978) Spectrochim Acta 33B: 551–576
6. Kaiser H, Specker H (1956) Fresenius Z Anal Chem 149: 46–66
7. Taylor JK (1983) Anal Chem 55: 600A-604A, 608A
8. BIPM, IEC, IFCC, ISO, IUPAC, IUPAP, OIML (1993) International vocabulary of basic and general terms in metrology, 2nd edn. International Organization for Standardization (ISO), Geneva
9. Wilson AL (1970) Talanta 17: 21–29
10. Doerffel K (1998) Fresenius J Anal Chem 361: 393–394
11. Shaevich AB (1989) Fresenius Z Anal Chem 335: 9–14
12. Terms, definitions, and symbols for metrological characteristics in analysis of substance (1975) Zh Anal Chim 30: 2058–2063 (in Russian)
13. Youden WJ (1960) Anal Chem 32 (13): 23A-37A
14. Youden WJ (1961) Mat Res Stand 1: 268–271
15. Zemelman MA (1991) Metrological foundations of technical measurements. Izdatelstvo standartov, Moscow (in Russian)
16. Eisenhart C (1963) J Res Nat Bur Stand 67C: 161–187
17. Campion PJ, Burns JE, Williams A (1973) A code of practice for the detailed statement of accuracy. National Physical Laboratory, Her Majesty's Stationery Office, London
18. BIPM, IEC, IFCC, ISO, IUPAC, IUPAP, OIML (1993) Guide to the expression of uncertainty in measurement. ISO, Geneva
19. EURACHEM/CITAC Guide (2000) Quantifying uncertainty in analytical measurement, 2nd edn (http://www.eurachem.bam.de/guides/quam2.pdf)
20. Ellison SLR, Barwick VJ (1998) Analyst 123: 1387–1392
21. Ramsey MH (1998) J Anal Atom Spectr 13: 97–104
22. van der Veen AMH, Alink A (1998) Accred Qual Assur 3: 20–26
23. Valcárcel M, Ríos A (1993) Anal Chem 65: 781A-787A
24. ISO 5725 (1994) Accuracy (trueness and precision) of measurement methods and results. Parts 1–6. International Organization for Standardization, Geneva
25. IUPAC (1995) Protocol for the design, conduct and interpretation of method-performance studies. Pure Appl Chem 67: 331–343
26. Makulov NA (1976) Zavod Lab 42: 1457–1464 (in Russian)
27. Parczewski A, Rokosz A (1978) Chem Analityczna 23: 225–230
28. EURACHEM (1998) The fitness for purpose of analytical methods. A laboratory guide to method validation and related topics. LGC, Teddington
29. Wegsheider W (1996) Validation of analytical methods. In: Günzler H (ed.) Accreditation and quality assurance in analytical chemistry. Springer, Berlin, etc., pp 135–158
30. Bruce P, Minkkinen P, Riekkola M-L (1998) Mikrochim Acta 128: 93–106
31. ISO/IEC 17025 (1999) General requirements for the competence of testing and calibration laboratories. International Organization for Standardization, Geneva
32. Jülicher B, Gowik P, Uhlig S (1999) Analyst 124: 537–545
33. van der Voet H, van Rhijn JA, van de Wiel HJ (1999) Anal Chim Acta 391: 159–171
34. GOST R 8.563–96 State system for ensuring the uniformity of measurements. Procedures of measurements. Gosstandart of Russia, Moscow (in Russian)

R. Kadis (✉)
D. I. Mendeleyev Institute for Metrology,
19 Moskovsky pr.,
198005 St. Petersburg, Russia
e-mail: rkadis@mail.rcom.ru
Tel.: +7-812-3239644
Fax: +7-812-3279776

Accred Qual Assur (2002) 7:299–304
DOI 10.1007/s00769-002-0494-7

A. Leclercq

Flexibilization of the scope of accreditation: an important asset for food and water microbiology laboratories

Abstract Beltest, the Belgian accreditation body, has investigated flexibilization of the scope of accreditation for chemistry laboratories and food and water microbiology laboratories. This flexibilization, synonymous with test-type accreditation, allows a laboratory to add new test methods or retry previous test methods without having to undergo a new audit by Beltest. It has been used for nearly ten years by German and Swiss accreditation bodies. Flexibilization permits the validation of methods and results, given that the competence of the particular laboratory is already well established. This new concept in microbiology allows client's needs to be adequately met, and facilitates the quick establishment of a method in several laboratories at once in case of a public health crisis. The first laboratory to participate at this investigation on the flexibilization concept, as a test of the concept, was the Belgian reference laboratory for food microbiology.

Keywords Accreditation · Food and water microbiology · Validation · Type of tests

Accreditation of Belgian laboratories is based on proof that they conform, as regards documentation and practices, to the requirements established in the standard guideline NBN-EN 45001 [1], and also, if necessary, to applicable requirements of sector-related or national accreditation agencies [2, 3, 4].

The NBN-EN 45001 standard has been replaced with the more precise standard NBN-EN-ISO-CEI 17025 [5], approved in December 1999 with a phase-in period of two years for application, which takes into consideration the importance of continuous quality assurance enhancement. In the same way, certification of quality systems based on standards of the NBN-EN-ISO 9000 series [6, 7], which evaluated only the capacity of laboratories to establish a quality control system, is changing over to standards based on the NBN-EN-ISO 9001 (Version 2000) standard [8], also taking continuous quality assurance enhancement into account. The process of development of the NBN-EN-ISO-CEI 17025 standard incorporated all the requirements of NBN-EN-ISO 9001 [6] and 9002 [7] that were relevant to the scope of testing services covered by the laboratory's quality system. These last requirements of NBN-EN-ISO 9001 and 9002 should be replaced by requirements of the NBN-EN-ISO 9001 (version 2000) standard during revision of the NBN-EN-ISO-CEI 17025 standard. Nevertheless, accreditation requires not only an evaluation of quality systems, but also an evaluation of the technical competence to perform specific tests.

The heart of the accreditation process is a scope of accreditation which is well-defined and which avoids ambiguity or multiple interpretations. Accreditation must constitute a credible attestation to the qualifications of a laboratory [4, 9, 10]. This scope of accreditation can affect in may important ways the development of the technical competence of a laboratory. At the same time, though, laboratories need to be able to use test methods that meet the growing needs of clients.

As regards food and water microbiology, current outbreaks in Europe have made apparent the necessity for official inspection agencies or ministry departments and for food industries or water agencies to get laboratories accredited quickly to perform tests in accordance with new test parameters, or to apply current methods to new ranges of food products. Accredited laboratories must be able to develop and validate new methods, or to derive methods from standards. Official recognition of these new test methods or their official incorporation into the scope of accreditation takes a long time, and audits are required for validation of these test methods.

A new concept "Flexibibilization of scope of accreditation" or "Accreditation of types of tests" or "Scope in testing" has emerged [11] in microbiology. It has been used by German and Swiss accreditation bodies for nearly ten years in other fields. Under this concept, an accredited laboratory, which has shown solid technical competence in the past, could, after establishment of a validation file, directly incorporate the new developed test method or application inside its scope of accreditation without a separate specific inspection. This laboratory could also at any moment retry new test methods or application. Surveillance audits then confirm (or fail to confirm) the applicability of the given test within the scope of accreditation. This is a modification of the requirements resulting from accreditation stipulated in paragraph 7, section b of the NBN-EN 45001 standard, and it is complementary to paragraph 1.6, 5.4.3 and 5.4.4. of the NBN-EN-ISO-CEI 17025 standard. As regards food and water laboratories, three principal headings were drawn up: formulation of scope of accreditation, flexibility with regard to food products tested, and validation of test methods and results.

Definition of scope of accreditation

Accreditation by individual tests

The area of application of an accreditation or scope of accreditation must be clearly defined by reference to one or more tests or types of tests and, if applicable, test items (NBN-EN-45002 [12]). Accreditation takes into account only precise individual tests as described with reference to the particular products analyzed, to determinate parameters, according to the test method that is prescribed, and also with reference to the technical procedure for carrying out a test or for implementing a particular standard (NBN-EN 45001).

Scope of accreditation comes to refer to a detailed list of individual tests in conformity with accreditation references and reflects the situation of a given laboratory at the time when an audit is conducted. Each test is specific to a product or type of products analyzed, to a set of parameters measured, to a range of measurement applicable to this particular test, and to test methods used by a given laboratory for this particular test. "Product" as a concept could also be expressed as a "range of product." Accreditation for a test method could drown methods considered as equivalent if they are established via identical methodology.

This approach is clear, strict, and comprehensible to all clients. Laboratories which perform the same functions over and over and which do not change the scope of their activity can by this means have their competence clearly established for potential clients. Due to its rigidity, any significant modification of the "scope of accreditation" concept would require an extensive audit by the appropriate accreditation agency.

The concept of "scope of accreditation" presents differences which can be categorized in various ways, with regard to customer satisfaction. Customer demands must be aligned with the real accreditation needs in scientific terms. Figure 1 illustrates the relationship between customer demand and real need, and the availability

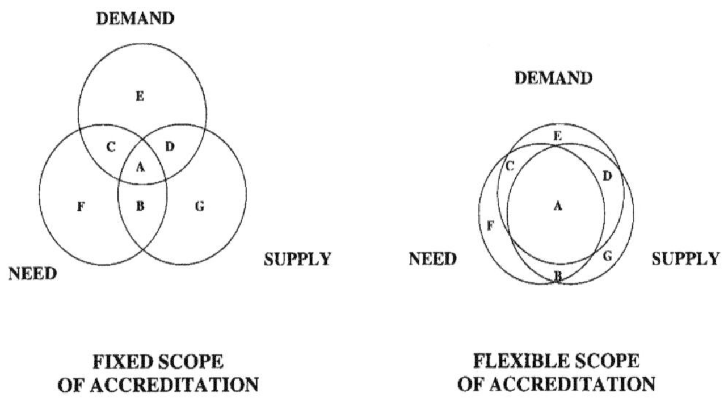

Fig. 1 Relationship between need, demand of clients and supply of laboratory for fixed and flexible scope of accreditation

(supply) of laboratory facilities, which can employ for fixed or flexible scope of accreditation. For fixed scope of accreditation, area A represents laboratory capacity in which quality is controlled; scope of accreditation is ideal. Area B represents capacity corresponding to the need and not to the demand; this is something that can improve customer satisfaction. Scope of accreditation offers new parameters for clients that could also offer them commercial advantages. This is an important part of the concept of flexible scope of accreditation. Area C is the part of the market where the new concept fails to meet specifications and so fails to satisfy customers. Scope of accreditation does not cover certain tests that may be required by some clients. Area D represents excessive quality control. Area E represents unnecessary or unsatisfied requirements; customers who request things they don't need. Area F represents shortcomings on the part of the accredited laboratory; improvements are necessary to satisfy potential needs. This is another important part of the concept of flexible scope of accreditation. Area G represents scope parameters that are unnecessary: sub-quality, possible gain for laboratory. Laboratories have a scope of accreditation constituted through analysis not necessary for their market. Under flexible scope of accreditation, categories of need, demand and supply tend to be confused: Area A becomes predominant.

Accreditation by types of tests or flexible scope

Recently, Beltest, the Belgian accreditation agency, investigated a new concept called "flexibilization of scope of accreditation" [11], which is an accreditation bearing on types of tests (i.e. enumeration of microorganisms) with reference to a test method (i.e. enumeration of *Clostridium perfringens*) and a test field (i.e. all foods, water), mainly associated with a product or a range of products. Flexibilization of the scope of accreditation is understood to comprise all accreditation procedures not directed exclusively at the accreditation of individual test methods [13]. A "test type" is defined as a group of tests characterized by utilization of similar methods (including sample preparation, standardization, calibration and validation principles, and training fundamentals) applied in a particular sector of a testing field. Types of tests may be defined on a technology/methodology-related or application-related basis. For each type of tests, the following must be specified: internal reference of laboratory, testing field(s) or products or range of products involved, test method(s) used and the eventual date of beginning and end of accreditation procedures in regard to the given test. Item(s) being tested could also be specified in some cases. A simplified example of flexible scope accreditation was given in Table 1. The definitions of flexible-scope accreditation will vary from country to country and from sector to sector depend-

Table 1 Example of presentation of flexible scope of accreditation

Internal code	Samples Test field (Products or materials or type of activity tested)	Measured principle Type of tests (Range of measurement, properties measured)	Description Test method (method, standard, validated in-house method)	Accredited from to
FLEX 1	**All foods and water, excepted worn water**	**Enumeration of microorganisms**[a]	**Culture**	
P20/I3.2.	All foods	Enumeration of total coliforms	NF V08-050[b]	
P20/I10.2.	All foods	Enumeration of *Clostridium perfringens*	NF V08-056	
FLEX 2	**Water, excepted worn water**	**Enumeration of microorganisms**[a]	**Culture after filtration**	
P20/I2.5	Water	Enumeration of presumed *Escherichia coli*	ISO 9308/1	
P20/I15.2	Water	Enumeration of *Pseudomonas aeruginosa*	NF-T-90-421	
FLEX 3	**All foods**	**Detection of microorganisms**[a]	**Culture**	
P20/ I 49.1	All foods	Detection of *Salmonella* spp	ISO 6579	
P20/I49.2	All foods	Detection of *Listeria monocytogenes*	ISO 11290-1	
FLEX 4	**All foods**	**Detection of microorganisms**[a]	**Immunodiagnostic**	
P20/I55.1	All foods	Detection of *Salmonella* spp	AFNOR-BIO-12/1-04/94 Vidas *Salmonella*	
P20/I55.3	All foods	Detection of *Escherichia coli O157*	AFNOR-DYN-16/1-03/96 Dynabeads anti *Escherichia coli* O157	

[a] Type of tests.
[b] NF: French Standard; ISO: International Organization for Standardization; AFNOR: French association for standardization; Bio: bioMérieux; DYN: Dynal

ing on established procedures in the various sectors and countries, and depending on the requirements and needs of important customers of the laboratory. This flexibility could take different forms, as indicated in Table 1:

1. Fixed scope with footnotes to measure principle column, which indicate that no modification of the list of accredited methods allowed and flexible scope with footnotes ("Optimization of given test methods allowed (adaptation to clients needs, new editions of test standards)", or, "Development of additional test methods within the accredited types of tests allowed") to measure principle column, which indicate flexibility levels after fixed scope.
2. Fixed scope parts are indicated by lists of accredited methods and flexible scope parts are indicated by reference to documented in-house methods and procedures.
3. Fixed scope parts are given by lists of test methods. Flexible parts are indicated by reference of type of tests. The rules of the accreditation body require the competence in a minimum number of testing methods for the accreditation of "types of tests".
4. Fixed scope parts are given by lists of test methods. Flexible scopes are given by short lists of standards and in-house methods followed by the sentence: "Within the indicated test area the laboratory may modify, improve and newly develop test methods without prior information and consent of the accreditation body. The test methods given are examples only" [14].

At all times, the laboratory being accredited must maintain a current listing of individual validations that can be made for a type of tests, available on request from the appropriate accreditation agency or its inspectors, and for external queries, i.e. from clients. Laboratories may introduce new test methods or modified methods, within the limits prescribed for the accredited type of tests, within the approved type of test, without the necessity to obtain approval from accreditation authority in each individual case. It might also be possible to obtain confirmation of competence with regard to non-routine tests on the basis of general procedures. A technical annex in accreditation certificate could refer simultaneously to individual tests and types of tests. Extension and/or a surveillance audit could integrate, definitively or not the new tests or type of tests validated by the laboratory. Information about modifications is provided at fixed surveillance intervals. All additions of new types of tests, testing fields and new test methods outside the accredited types of tests in the scope of accreditation must be reported to the accreditation authority as part of an extension.

NBN-EN-ISO-CEI 17025 standard states in its paragraph 1.6 that "*If testing and calibration laboratories comply with the requirements of this International Standard they will operate a quality system for their testing and calibration activities that also meets the requirements of ISO 9001 when they engage in the design/development of new methods, and/or develop test programs combining standard and non-standard test and calibration methods, and ISO 9002 when they only use standard methods*" [5]. This correspondence may be mentioned in the accreditation certificate or in its relevant annexes.

So, the meaning of "flexibility of scope of accreditation" is a group of measures takes to not limit accreditation at individual tests. It addresses testing proficiency and is not generally intended for research and/or product development laboratories, unless specified by a client and/or proficiency to a test method is critical to these functions.

Requirements and assessment of laboratory

Requirements: accreditation body aspect

A laboratory may be eligible for flexible scope of accreditation or accreditation by type of tests only if it is accredited for individual tests and if it has passed at least one surveillance audit. Flexibility might be applied on methods of tests and/or types of analyzed products.

Accreditation by type of tests must satisfy particular requirements defined by a Beltest guide [11] and a demonstration by the given laboratory of its performance with regard to each type of tests. For each type of tests, the performance of the given laboratory in the following respects must be submitted to an appropriate accreditation agency:

1. A sufficient number of different individual tests, at least six, concerning type of tests and corresponding validation or verification reports (in certain exceptional cases these could be routine tests, but in such a case a supplemental validation or verification procedure must be carried out);
2. General procedures of validation or verification applicable to type of test concerned;
3. Records for the corresponding validation or verification.

A list of the test methods currently covered by the existing accreditation must be available at all times in the testing laboratory on request of accreditation body. As part of the monitoring procedure, this list must be submitted to the accreditation authority with the new or modified test method identified.

Requirements: laboratory aspect

For a food or water laboratory, each type of test which must be accredited must satisfy several types of requirements:

1. Each test included in a given type of tests must be submitted to a validation procedure applied to the implementation of the methods in the laboratory, and must be regularly evaluated with triply redundant verification procedures;
2. When laboratories use outdated versions of standards, or fail to mention a date of version of standards in their statement of accreditation scope, they must revise their procedures for testing (revise instructions) and train technicians in new procedures for testing according to new standards within a fixed short period after standards publication;
3. When the flexible-scope accreditation bears on methods for "all foods" or "all waters", laboratories must be under strict control during validation of test results and at reception of samples. Test results depend completely upon sampling carried out by the laboratory, and on the first step of manipulation of sample before real test application. Validation of implementation of an accredited test method is made in accordance with a divergent type of matrix representing the main activities of the given laboratory. The laboratory must establish and scientifically document the proper handling of all the different types of food samples delivered by clients. Upon reception of a non-conventional sample, technicians must have definite procedural instructions for handling the sample. This step is necessary for the correct identification of the sample. Also, the availability of the sample material and the competence of the technician must be verified by an appropriate accrediting agency.

Assessment

Several points which arise during the preparation of accreditation files in accordance with this new concept, or which arise during inspection, must be reviewed:

1. Organizational aspects that the laboratory must monitor in relation to validation or verification of new or modified test methods;
2. Competence, experience and training (present and future) of staff (and particularly for senior technical and management staff) in statistics and in relation to the particular type of testing, and the particular products concerned;

3. Performance level of equipment;
4. Test procedures or/and instructions;
5. Performance level of quality control system;
6. Recording of completed validation procedures;
7. Conformation level by comparison of accreditation requirements, as mentioned by results of most recent audits (and particularly control of technical aspects);
8. Results of participation in ring-tests or collaborative trials;
9. Management specifications in cases of less frequent testing.

Inspection by an official accreditation agency must be based on a selection of certain test methods from both quantitative and qualitative points of view. This selection takes into consideration:

1. Evidence of the implementation of the quality management system, experience, capability, if any, modification/development of testing methods;
2. The technical complexity of the tests;
3. The possible consequences of errors (possible risks) in the tests;
4. The frequency of use of the test methods;
5. The ratio of routine tests (standard methods) to non-routine tests (special specifications from clients, in-house methods, ...);
6. The ratio of complete observations of test performance and checks of test reports to inspections of test facilities.

The range of audited activities could make it possible to attest the capacity of a given laboratory to introduce new test methods or to modify current methods. However, the testing laboratory must not be made to incur unnecessary costs.

Validation of methods and results: main point of flexible schedule

Laboratories which conform to standards NBN-EN 45001 (Paragraph 5.4.1., 5.4.2., 6.2 and 6.3.) and NBN-EN-ISO-CEI 17025 (Paragraph 5.4.5.) standards should use validated methods for testing, methods which produce reliable results; part of the results validation procedure includes an obligation to demonstrate the capacity to carry out this procedure.

Validation of methods

Validation is defined as confirmation by examination, including the provision of objective evidence that the particular requirements for a specific intended use are fulfilled. It may include procedures of sampling, handling and/or transportation. Procedures themselves, and any assigning of tasks such as development, implementation and validation of new or in-house test methods, should be laid out in detail, by a person experienced in documentation of quality control, in a flow chart format. For complex methods such procedures may, at least to some extent, end up in project management plans.

This validation procedure must confirm that the method under examination is adapted to the test context, and is continuously in accordance with the client's requirements or needs, as regards the following parameters: trueness, precision (reproducibility and/or repeatability), uncertainty of the results, measurement range comprising limits of detection and/or quantification, specificity and robustness against external influences and/or cross-sensitivity against interference from the matrix of the sample. Qualitative methods must satisfy the parameters of the matrix effect (and interference effect) which are very important when a flexible scope of accreditation is applied, because of the type of tests: that is, when "all foods" must be controlled. Levels of validation are evaluated depending on the particular type of method used, and on results. But the following parameters must be assumed:

1. Standardized methods of testing (and methods validated by an official agency of validation), used strictly (without adaptation of test parameters), limited to specified tasks (type of products tested, range of measurement) not requiring individual validation;
2. Modifications of standardized methods of testing (modification of test parameters, and/or specified tasks validated in regard to the modified parameters);
3. Non-standard methods (published methods, new test methods, commercialized test systems, kits, ...) or laboratory-designed/developed methods must be thoroughly validated.

The laboratory shall record the results [15] obtained in validation testing, the procedure used for validation, and a statement as to whether the method is appropriate for the intended use.

Certain official methods which are not considered in this list present some problems as regards validation. These frequently occur because of a failure to use standardized methods, or from simplification of standardized methods, sanctioned by an expert consensus, or written in to certain regulations. Their official appearance seems to influence accreditation bureaus to consider standardized and official methods of validation as equivalent. In many cases, these official methods were originally established in emergency situations of national crisis (outbreak, epidemic, etc.). In the majority of cases, these official methods lack validation procedures. Onerous standardized validation procedures may be circumvented. In such cases laboratories must demonstrate at a minimum the qualitative recovery of a microorganism for enrichment procedures which are replicated, and demonstrate quantitative recovery of a microorganism for an enumeration procedure carried out in replicate.

Validation of testing methods always involves a balance between costs, risks and the technical capabilities of a given laboratory (NBN-EN-ISO-CEI 17025, 1999). Laboratory administrators must establish minimum quality requirements before starting the process of validation and implementation, or before starting the whole development process. Procedures of validation for a small laboratory could not be the same as for commercial validation of a new method, whether or not new kits from a manufacturer are used. There are many cases in which the range and uncertainty of the test result values can only be given in a simplified way due to lack of information.

Validation and verification of results

Results verification is totally different from results validation. Results validation (point 4.7.5. and 5.9. of NBN-EN-ISO-CEI 17025 standard) shows, each year, or when it is judged necessary, that a given laboratory has the capacity to apply a particular method, repetitively, in respect of obtained data during initial validation. Trueness and statistical dispersion of results are the basis of the definition of the uncertainty of the standard of measurement [16] and, in some cases, the basis for the definition of the limit of detection and quantification. Management of data from validation results, as control card, could permit the detection and control of eventual deviation. Validation of results is the internal quality control procedure which verifies the stability of performance of the methods for which accreditation is sought, in the limited-scope procedural context.

Guideline for a validation scheme

Table 2 shows requirements for validation of quantitative methods and could be simplified in the case of qualitative methods.

So, the methods characteristic of each test, comprising taken together a type of tests, must undergo validation testing of their results. This is the implementation of the method, and the establishment of a standard for its performance. For the standardization of quantitative methods, this consists at a minimum of a determination of trueness when blank utilization, certified reference materials (or reference materials, or spiking materials) or collaborative trials are used, repeatability (*r*) with repetition,

Table 2 Validation parameters required for quantitative or qualitative test method

Methods	True ness	Repeat-ability	Intralaboratory reproducibility	Limit of detection	Limit of quantitation	Relative accuracy official to methods	Matrix effect	Linearity	Selectivity	Specificity	Sensibility	Robustness
New method	+[a]	+	+	(+)[b]	(+)	(+)	(+)	(+)	+	+	+	+
Classic method	+	+	+	(+)	(+)	+	(+)	–[c]	–	–	–	–
Modified method	+	+	+	(+)	(+)	(+)	(+)	(+)	(+)	(+)	(+)	(+)

[a] Always executed.
[b] Executed if technically justified.
[c] Not executed

or reference materials utilization and reproducibility (R) with collaborative trials or certified reference material (or reference materials). For the standardization of qualitative methods, the minimum requisite is the determination of limits of detection, with analysis by checked samples near this limit. These results must be reported, based on a significant number of tests, and discussed in a validation report, which give positive agreement to include new methods of testing as part of an accreditation program.

At the same time, internal quality control must be carried out to verify the performance stability of the limited-scope performance of the method. Triply redundant verification methods are carried out with regard to control of first, second and/or third line, each set of methods applied to a particular type of test. The first line of verification involves pro forma repetition of all the steps of the test, in order to establish repeatability or reproducibility for quantitative tests, and to verify the range of sensitivity or detection for qualitative tests. This verification is performed by the experimenter himself, as part of the proper performance of the test. A second line of verification is put into operation by administrative decision, and includes testing with "blind" samples, repetition of samples, internal audit procedures, etc. If necessary, a third line of verification can be set up by the use of certified reference materials (or spiking materials), or through collaborative trials. These procedures are based on external cooperation. External audit procedures and complaints handling procedures are also part of this third line of verification.

An annual schedule of validation of methods for each type of test could be established as regards limited-scope accreditation. Some test methods similar among themselves may allow the use of previously established validation procedures, such as those used for reference testing, or those used for the establishment of routine standards for the enumeration of aerobic mesophilic microorganisms. The schedule could differentiate between methods that can be validated during collaborative trials in which the laboratory participates, and methods where substrates must be used as reference materials if available.

Conclusion

Accreditation with flexible scope allows laboratories who receive requests for evaluation and analysis to quickly provide customers with a service adapted to their actual needs (Fig. 2), and provides food and water microbiology authorities with new test methods which can be brought rapidly to bear in cases of public health crisis.

Its implementation is easy in small countries due to centralization of management and ease of surveillance of this new form of accreditation by a national accreditation body. Due to the long-standing establishment of validation of food microbiological methods in some countries, the beginning of validation of water microbiological methods, and establishment of European standards for the validation of food microbiological test methods (Microval project), the flexible-scope type of accreditation must be adapted to certain national requirements, but could prove less complex than the standard requirements.

Acknowledgements I thank Doctor Yasmine Ghafir (Université Libre de Liège, Liège) and Nicole Vanlaethem (Beltest, Brussels) for their help.

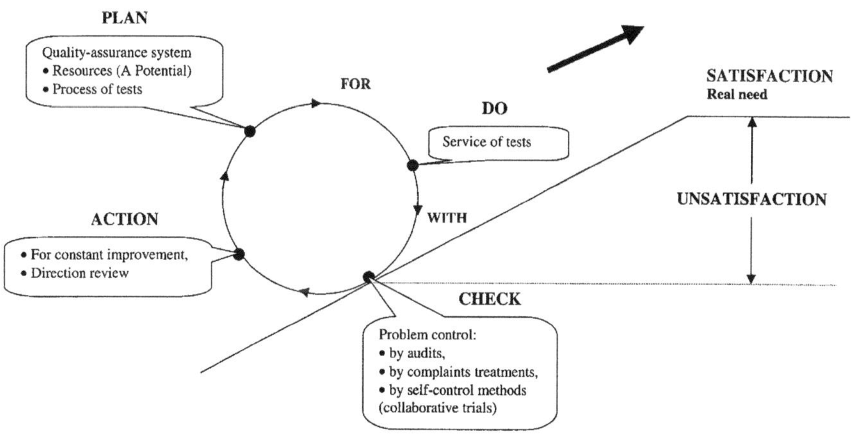

Fig. 2 Testing laboratory, accreditation requirements and relation clients-laboratory

References

1. CEN (1989a) NBN-EN 45001 – General criteria for the operation of testing laboratories. CEN, Brussels
2. Beltest (1991) Critères généraux et lignes directrices à l'usage des laboratoires candidats à une accréditation. Document L1/F-REV0. Beltest, Brussels
3. Eurachem (1996) Accreditation for laboratories performing microbiological testing. European cooperation for accreditation of laboratories. EAL-G18. Eurachem
4. Wilson B, Weir G (1995) Food and drink laboratory accreditation, a practical approach. Chapman & Hall, London
5. ISO (1999) NBN-EN-ISO-CEI 17025 – General requirements for the competence of testing and calibration laboratories. ISO, Geneva
6. ISO (1994) NBN-EN-ISO 9001 – Quality Systems – Model for quality assurance in design/development, production, installation and servicing. ISO, Geneva
7. ISO (1994) NBN-EN-ISO 9002 – Quality Systems – Model for quality assurance in production, installation and servicing. ISO, Geneva
8. ISO (2000) NBN-EN-ISO 9001:2000 – Quality management systems – Requirements. ISO, Geneva
9. Flowers RS, Gecan JS, Pusch DJ (1992) Laboratory quality assurance. In: Vanderzant C, Splittstoesser DF (eds) Compendium of methods for the microbiological examination of foods. APHA, Washington, DC
10. Lightfoot NF, Maier EA (1998) Microbiological analysis of food and water. Guidelines for quality assurance. Elsevier, Amsterdam
11. Beltest (2000) I015 – REV.0 – Critères auxquels doivent répondre les laboratoires accrédités demandeurs d'un domaine d'application flexible pour les analyses de pesticides. Beltest, Brussels
12. CEN (1989b) NBN-EN 45002 – General criteria for the assessment of testing laboratories. CEN, Brussels
13. Steinhorst A (1998) Accreditation and Quality Assurance 3:295
14. ILAC (2000) Draft 6 – The scope of accreditation and consideration of methods and criteria for the assessment of the scope in testing. ILAC, Rhodes, Australia
15. Beltest (1996) Lignes directrices relatives à la traçabilité des mesures. Document L8/REV0. Beltest, Brussels
16. ISO (1994) ISO 5725-1. Exactitude (justesse et fidélité) des résultats et méthodes de mesure – Partie 1: Principes généraux et définitions. ISO, Geneva

A. Leclercq (✉)
Institut Pasteur de Lille,
Department of Food Microbiology and Safety,
Rue du Professeur Calmette 1,
59000 Lille, France
e-mail:
Alexandre.Leclercq@pasteur-lille.fr
Tel.: +33-3-20-43-89-23
Fax: +33-3-20-43-89-26

Relevant legislation, legislation plans and related initiatives in different countries and continents

Accred Qual Assur 3:211–214

H. Hey

Different approaches to legal requirements on validation of test methods for official control of foodstuffs and of veterinary drug residues and element contaminants in the EC

Abstract In order to ensure food consumer protection as well as to avoid barriers to trade and unnecessary duplications of laboratory tests and to gain mutual recognition of results of analyses, the quality of laboratories and test results has to be guaranteed. For this purpose, the EC Council and the Commission have introduced provisions

- on measures for quality assurance for official laboratories concerning the analyses of foodstuffs on the one hand and animals and fresh meat on the other,
- on the validation of test methods to obtain results of sufficient accuracy.

This article deals with legal requirements in the European Union on basic principles of laboratory quality assurance for official notification to the EC Commission and on method validation concerning official laboratories. Widespread discussions and activities on measurement uncertainty are in progress, and the European validation standards for official purposes may serve as a basis for world-wide efforts on quality harmonization of analytical results. Although much time has already been spent, definitions and requirements have to be revised and further additions have to be made.

Key words Laboratories for offical control of foodstuffs, veterinary drug residues and element contaminants · Validation of test methods · Legal requirements

Laboratories for official control of foodstuffs, veterinary drug residues and element contaminants

Official laboratories testing foodstuffs and stockfarming products of animal origin have to meet requirements of two separate and distinct legislative areas including analytical quality assurance in the EC. The first area concerns the official control of foodstuffs. It falls within the remit of the Directorate General III (DG III), Internal Market, and refers to Directives 89/397/EEC [1] and 93/99/EEC [2]. The second area is within the responsibility of the Directorate General VI (DG VI), Agriculture, and refers to a series of directives which until May 1996 regulated the prohibition of the use of certain substances having a pharmacological action and which are used in stockfarming to promote growth and productivity in lifestock or for therapeutic purposes. In order to harmonize the legal regulations of the past on residues of illegal substances, a number of provisions required clarification in the interests of the effective application of control and residue detection in the Community. With a view to ensure the immediate and uniform application of the official controls, the rules of the past and their amendments were assembled in the Councils Directive 96/22/EC [3].

It is the purpose of the official food control, of the veterinary drug control and of the element contaminant control to pursuit the free movement of foodstuffs, living animals and fresh meat within the EU and to guarantee safe products without health hazards. Additionally, the official food control shall prevent fraud from consumers, whereas a main task of veterinary drug control (official and self control) is to compel the stockfarming industry to take greater responsibility for the quality and safety of meat and to guarantee even odds. This is the reason why Art. 31 of 96/23/EC [4] provides for the charging of a fee to cover the official control measures on drug and element contaminant analysis.

Unfortunately; DG III and DG VI are following totally different approaches to legislation, laboratory quality assurance and test method or test result validation:

The official status of laboratories in food control is defined in Art. 7 of the Council Directive 89/397/EEC, which refers to both "first opinion" and "second opinion" (or referee) laboratories. "First opinion" stands for laboratories of the official food control authorities in the European Union. "Second opinion" means private third party laboratories which undertake analyses on a "second part of samples" from such enterprises, which have been subject to official inspection and sampling by the enforcement authorities. The status is laid down in the basic food law of some EU member states, and the laboratories have to be officially acknowledged to the EC Commission not later than 31 October 1998. Their function is to enable food companies to defend their legitimate claims against the food enforcement authorities. The private laboratory community is not quite familiar with this clarification of the EU Commission and the EU Council in the Council minutes [6] on Art. 3 of the Council Directive 93/99/EEC.

First and second opinion institutions for the purpose of official food analysis have to be notified to the Commission as laboratories referred to in Art. 7 of the Directive 89/397/EEC. They have to meet the assessment criteria laid down in Art. 3 § 1 and § 2 of the Directive 93/99/EEC. The regulations consist of the general criteria for the operation of testing laboratories laid down in European Standard EN 45001 supplemented by standard operating procedures and the random audit of their compliance by quality assurance personnel. These supplements are in accordance with the OECD principles nos. 2 and 7 of good laboratory practice as set out in Section II of Annex 2 of the Decision of the Council of the OECD of 12 May 1981 concerning the mutual acceptance of data in the assessment of chemicals [7]. The measures and administrative provisions necessary to comply with this legal requirement have to be met before 1 November 1998. The accreditation pursuing EN 45001 in the valid status alone does not meet the supplements (standard operating procedures and random audit by quality assurance personnel, which is not involved in testing) as legal requirements of official food and second opinion laboratories as mentioned above.

Concerning laboratories for the analysis of veterinary drug residues and element contaminants in animals and fresh meat, there are no detailed regulations on the assessment of laboratory qualitiy assurance in contrast to official food control laboratories, but Annex V, Chapter 2, no. 1b) of the Council Directive 96/23/EC refers to functions and activities of Community reference laboratories in this field. Function no. 1b) includes the support of the national residue reference laboratories to build up an appropriate system of quality assurance, which is based on the general criteria of good laboratory practice (GLP principles) and the standards series EN 45000. This system can be regarded as analogous to the legal requirements laid down for official food control laboratories in Art. 3 of Directive 93/99/EEC. So far, in contrast to official food laboratories, there are no legally binding provisions for the assessment and notification of official veterinary drug- and element-contaminant laboratories to the EC Commission. But if the acknowledgement of these laboratories is to become mandatory in the future, the basic system of quality assurance seems to be quite clear: it will consist of some elements of the OECD principles of GLP and of the standard series EN 45000. Besides the general principles of laboratory organization and quality assurance there are existing detailed legal EC regulations on validation of methods referring to both kinds of laboratories.

	DG III, Internal Market official food control	DG VI, Agriculture veterinary drug and element contaminant control
Legislation	Preferably horizontal	Preferably vertical
Laboratory quality assurance system for notification to the EC-Commission	Yes, Art. 3 of Directive 93/99/EEC: EN 45000 series, additional measures of OECD-GLP principles	No
Methods of analysis	Preferably optional to the laboratories, regulations on validation	Preferably method standardisation, additional regulations on validation (going further than DG III)
Reference laboratory system	No	Yes
Regulations on proficiency testing	Yes, Art. 3 of Directive 93/99/EEC, International Harmonized Protocol [5] (according to Council minutes [6])	No (indirect: collaborative trials, managed by reference laboratories)

Validation requirements concerning offical food control laboratories

Art. 4 of the Directive 93/99/EEC requires the member states of the EU to ensure the validation of methods of analysis used within the context of official control of foodstuffs "whenever possible". For this purpose, the laboratories have to comply with the provisions of paragraphs 1 and 2 of the Annex to Council Directive 85/591/EEC [8] concerning the introduction of Community methods of sampling and analysis for the monitoring of foodstuffs intended for human consumption.

Annex 1 and 2 in detail demand:

1. Methods of analysis which are to be considered for adoption under the provisions of the Directive shall be examined with respect to the following criteria:

(a) specificity,
(b) accuracy,

(c) precision; repeatability intra-laboratory (within laboratory) and reproducibility interlaboratory (within and between laboratories) variabilities,
(d) limit of detection,
(e) sensitivity, (which means the slope of the calibration curve and which is often mixed up with the level of the detection limit)
(f) practicability and applicability,
(g) other criteria which may be selected as required.

2. The precision values referred to in 1 (c) shall be obtained from a collaborative trial which has been conducted in accordance with an internationally recognized protocol on collaborative trials [e.g. International Organization of Standarization "Precision of Test Methods" (ISO 5725/1981)]. The repeatability and reproducibility values shall be expressed in an internationally recognized form (e.g. the 95% confidence intervals as defined by ISO 5725/1981). The results from the collaborative trial shall be published or freely available.

Validation requirements concerning official control laboratories on veterinary drug residues and element contaminants

Validation criteria referring to official laboratories on the control of veterinary drug residues are included in the following Commission Decisions:

- 93/256/EEC [9], which lays down the methods to be used for detecting residues of substances having a hormonal or a thyreostatic action (immunoassay, thin-layer chromatography, liquid chromatography, gas chromatography, mass spectrometry, spectrometry or other methods),
- 90/515/EEC [10], which lays down reference methods to be used for detecting heavy metals and arsenic.

Commission Decisions are strictly mandatory to those laboratories to whom they are addressed. Compared to a Commission Decision, a Council or Commission Directive has not the same legally binding character, but the general aims of a Directive have to be transposed into the member states' national law.

Whereas the Decision 93/256/EEC itself is very brief, the annex includes

- definitions of terms related to analyses,
- general requirements, which in accordance with Directive 85/591/EEC shall apply to the examination of test methods, and
- criteria for the qualitative identification and quantification of residues.

In detail, the Decision 93/256/EEC in the first part of the annex demands the same validation criteria as the Directive 85/591/EEC, but beyond this there are the following additional requirements:

- sample blank determination
- reagent blank determination
- requirements for calibration curves
- limit of determination (additional to limit of detection in 85/591/EEC)
- co-chromatography (second run of analytical procedure with standard addition)
- calibration curves
- susceptibility to interference.

Pursuing Art. 3 of 93/256/EEC, the validation criteria laid down in the Annex "are applicable to routine methods of analysis of residues of substances having a hormonal or a thyreostatic action". The Annex of 93/256/EEC distinguishes between "screening methods" (no. 1.2.2) and "confirmation methods" (no. 1.2.3). According to Art. 3 of the Commissions Decision 93/256/EEC, the calibration criteria for confirmation methods are applicable to reference analyses on veterinary agents, which are laid down in Annex I of Directive 86/469/EEC [11] and which refer to the control of animals and fresh meat.

In comparison to the Directive 85/591/EEC and to the Decision 93/256/EEC the Decision 90/515/EEC on the analysis of element contaminants does not contain further basic validation criteria. Both 93/256/EEC and 90/515/EEC demand that special requirements for the interpretation of results are fulfilled in order to ensure accurate results and/or to prevent false positive or negative signals.

Discussion

The Annex of 93/256/EEC is an almost complete guideline on the validation of physico-chemical methods of analysis. Although the title concerns the analysis of hormonal and thyreostatic agents in animals and fresh meat, the criteria should be applied to the routine testing of any analyte irrespective of the kind and origin of samples and matrices. They should neither be restricted to substances having a hormonal or a thyreostatic action nor to reference methods for official purposes nor to laboratories for the purpose of official enforcement authorities. Besides the official laboratories, the private ones, which are working in the field of food, of veterinary drug and of element contaminant analyses, should focus their attention on the validation criteria mentioned above as well.

Since no guideline may be perfect, the Annex to 93/256/EEC needs to be revised because

1. Some definitions are contradictory, meaningless, without benefit or will cause much expenditure of personnel and measurement capacity, e.g. "Limit of determination. This is the smallest analyte content for which the method has been validated with specific accuracy and precision". Apart from the fact that "precision" is included in the explanation of "accuracy"; the definition manifests a fundamental inability to give a definition which is fit for practice. A useful definition of the detection and quantification limit is based on a statistical approach to the confidence hyperbola of a methods calibration curve, elaborated by the "Deutsche Forschungsgemeinschaft" [12].
2. Some requirements cannot be fulfilled in routine analyses, because of being too costly, time expensive and large-scaled, e.g. general requirements on sample blanks and calibration curves, where among others the following procedures have to be followed and/or information must be given:

- In case of nonlinearity, the mathematical formula which describes the curve.
- Acceptable ranges within which the parameters of the calibration curve may vary from day to day.
- Verification of the calibration curve by internal standards.
- A minimum of six calibration points is required.
- Details of the variance of the variable of the calibration curve should be given.
- Control samples have to be included in each assay, with concentration levels zero and at lower, middle and upper parts of the working range.
- The procedure of analysis shall include sample blank and reagent blank determinations.
- If it is to be expected that factors such as species, sex, age, feeding or other environmental factors may influence the charateristics of a method, a set of at least 20 blank samples is required for each individual homogeneons population to which the method is to be applied.

These requirements seem to be "hatched at the green table", because, if they were to be fulfilled completely, the "practicability" requirements of the Commission Decision such as high sample throughput and low cost could never be met.

The Decision 93/256/EEC in case of repeated analyses of a reference material includes the following ranges for deviation of the mean from the true value (accuracy) and coefficients of variation

True value (mass fraction)	Deviation ranges
<1 μg/kg	–50% to +20%
>1 μg/kg – 10 μg/kg	–30% to +10%
>10 μg/kg	–20% to +10%

Content (mass fraction)	Coefficient of variation
1 μg/kg	45%
10 μg/kg	32%
100 μg/kg	23%
1 mg/kg	16%

These deviation ranges concerning accuracy and precision should be completed by ranges for the recovery, e.g. in accordance with a report of the Netherlands Inspectorate of Health Protection on validation of methods [13]:

Analyte content	Minimum recovery
>0,5 mg/kg	80–110%
1 to 500 μg/kg	60–115%
<1 μ/kg	50%

Furthermore, there should be minimum requirements on the use of certified standard reference samples and the use of proficiency testing schemes as far as appropriate. Pursuing Art 3 § 2b of the Council Directive 93/99/EEC it is mandatory to official food laboratories to carry out proficiency tests, which according to the "statement for the Council minutes"[6] have to comply with the International Harmonized Protocol for the Proficiency of (Chemical) Analytical Laboratories [5].

Conclusion

Harmonization of requirements on analytical quality assurance and method validation between DG III and DG VI is urgent and should have absolute priority. The validation criteria from 93/256/EEC described and discussed above are partially mandatory to official EU laboratories for the analysis of foodstuffs and have to be fully met by official laboratories testing veterinary drug residues in animals and fresh meat. The examples given do not claim to be complete. They may serve as a guideline, and there are many good reasons for a general revision and for adding further requirements. Besides, the guideline may be a good basis for worldwide harmonization of physicochemical method validation, particularly in the current discussion on measurement uncertainty.

References

1. OJ No L 186/23 of 30.06.1989
2. OJ No L 290/14 of 24.11.1993
3. OJ No L 125/3 of 23.05.1996
4. OJ No L 125/10 of 23.05.1996
5. Thompson M, Wood R (1993) J AOAC Int 76:926–940
6. Document 4241/94/EEC, statement for the Council minutes
7. Good Laboratory Practice in the Testing of Chemicals, OECD, Paris 1982
8. OJ No L 372/50 of 31.12.1985
9. OJ No L 118/64 of 14.05.1993
10. OJ No L 286/33 of 18.10.1990
11. OJ No L 275/36 of 26.09.1986
12. Rückstandsanalytik von Pflanzenschutzmitteln, Deutsche Forschungsgemeinschaft, 11th edn, 1991
13. Report number IGB/93-002, second version, Rijswijk, March 1993

H. Hey (✉)
Official food and veterinary laboratory of Schleswig-Holstein, D-24537 Neumünster, Germany
Tel.: +49-4321-904610
Fax: +49-4321-904619

Accred Qual Assur (1998) 3:32–35

Margreet Lauwaars

Methods validation: AOAC's three validation systems

Abstract Validated methods of analysis are needed for many purposes: enforcement of regulations, import/export control, in accredited laboratories, academia, institutions. The AOAC INTERNATIONAL Official Methods Program is designed to provide fully validated methods of analysis, based on interlaboratory testing by a minimum of eight laboratories. Another, lesser validation system is used for peer-verified methods of analysis where two or three laboratories participate. The system for performance testing of test kits is specially designed for a thorough testing of manufacturer claims, and can be obtained by submitting a kit to Performance Testing by the AOAC Research Institute.

Key words Validation of methods of analysis · Interlaboratory study · Peer-verified methods · Performance testing of test kits

Introduction

Since its beginning in 1884, AOAC INTERNATIONAL has been truly dedicated to the validation of analytical methods through trials in multiple laboratories. An early undertaking of AOAC is still its most important business: supporting the use of analytical methods used in multiple laboratories through validation by interlaboratory studies.

What do validated methods mean to the laboratory? Why use methods that have been performance-characterized through interlaboratory study validation? Such methods give the analyst increased confidence in the results of this analysis. In addition, the user of analytical data, who frequently may not be the analyst or the laboratory, will have increased confidence in the results of the analysis. This is important since regulatory actions or transactions from local to world level can be based upon these results.

The characteristics of the analytical method and its applicability to the work at hand are established through validation, are key elements in understanding the quality of analytical data, and assure confidence in quality.

> Why quality measurements are needed:
> - compliance with regulations
> - maintain quality and process control
> - meet terms of procurement
> - make regulatory decisions
> - conduct national and international trade
> - support research

Quality measurements have several elements. Quality assurance plans and quality control procedures are an essential beginning. In addition, it is necessary to have qualified scientists whose training needs to be documented and updated on a continuous basis. Quality measurements also require proper use of reference materials where available, and laboratories must repeatedly test their ability to perform through taking part in proficiency testing schemes. The provision of another essential element in quality measurements, namely validated methods, is the primary contribution from the work of AOAC.

There are many reasons why quality measurements are needed. Products must be tested for compliance and content, for safety, and to meet a variety of regulations. Quality measurements are also needed to monitor quality during production and in process control. Contracts are written based on standards and terms of procurement that must be tested. Every day, regulatory agencies make decisions based on results which are only as good as the quality of the methods upon which they themselves are based. Quality measurements support international and national trade and support research. Laboratories need to generate credible data and to do so they must have valid methods.

> Validation of a method establishes through systematic laboratory studies, that the performance characteristics of the method meet the specifications related to the intended use of the analytical results. Performance characteristics determined include: selectivity and specificity, range, linearity, sensitivity, limit of detection and limit of quantification, accuracy and precision.

Three method validation systems are operated by AOAC INTERNATIONAL:
- The AOAC Official Methods Program
- The AOAC Peer-Verified Methods Program
- The AOAC Performance-Tested Test Kit Program.

The AOAC Official Methods Program

The AOAC Official Methods Program is designed to provide analytical methods for which the performance characteristics have been validated to the highest degree of confidence recognized by an internationally harmonized protocol through an independent multiple laboratory study [1].

An acceptable interlaboratory study should be designed to support the intended scope of the method and to validate the performance across that scope. The number and type of materials sent to the collaborators is very important: Material is defined as a matrix/analyte combination, the aim being to have five matrices (or five concentrations) for each analyte. In the case of multianalyte methods, more than one analyte may be present in the same matrix, as a way of reducing the total number of matrices sent to collaborators. Each of the individual analytes must be present in those five matrices, as a minimum (i.e. one cannot send one matrix containing five analytes and consider that to be five materials).

For quantitative methods at least five materials must be included, a requirement which can be met by a combination of samples and levels of analyte. At least eight laboratories must take part and follow the interlaboratory study protocol. For reason of possible outliers, it is recommended that more than this number of laboratories should take part.

AOAC requires the following parameters to be included in the study for quantitative methods:
1. Reproducibility coefficient of variation RSD_R (CVR) at the appropriate concentration level
2. Repeatability (within laboratory) standard deviation, Sr
3. Reproducibility (including repeatability) standard deviation, S_R
4. % recovery
5. Method applicability statement
6. RSD_r(CVr), repeatability coefficient of variation
7. Repeatability, r (= 2.8 sr); and reproducibility R(= 2.8 S_R)
8. Mean
9. Number of laboratories
10. Number of outliers.

For qualitative methods providing a yes or no result, it is necessary to have analyte concentrations at two levels for each matrix. Each matrix must be represented by five samples. Likewise, five negative controls are required for each matrix. In total there would be no less than 15 samples in a qualitative study, and 15 laboratories must participate in this type of interlaboratory study.

AOAC requires the following parameters for qualitative methods:
1. Sensitivity rate
2. False negative rate
3. % agreement
4. Specificity rate (for non-microbiological methods)
5. False positive rate (for non-microbiological methods)
6. Number of laboratories
7. Number of outliers
8. Level of analyte (if known).

Since microbiology studies are treated in AOAC as qualitative studies, the large number of participating laboratories may cause problems. The AOAC Official Methods Board has on its agenda the development of a separate protocol for microbiology studies.

Ten steps to an AOAC Official Method:
1. Topic proposal to AOAC Office
2. Volunteer to conduct in-house validation
3. Review of in-house data and protocol
4. Recommendation for interlaboratory study
5. Recruit laboratories; conduct interlaboratory study
6. Analyse results; write report
7. Review results
8. Recommendation for adoption
9. Review by Official Methods Board (OMB)
10. OMB vote for adoption as First Action

Table 1 Data requirements of the AOAC Method Validation Programs

AOAC Official Methods	AOAC Performance-Tested Methods	AOAC Peer-Verified Methods
1. Specificity	1. Specificity	1. Specificity
2. Sensitivity	2. Sensitivity	2. Sensitivity
3. False positives/negatives	3. False positives/negatives	3. False positives/negatives
4. Limit of detection	4. Limit of detection	4. Limit of detection
5. Precision	5. Precision	5. Precision
6. Accuracy	6. Accuracy	6. Accuracy
7. Matrixes	7. Matrixes	7. Matrixes
8. Ruggedness	8. Ruggedness	8. Ruggedness
9. Recovery	9. Recovery	9. Recovery
10. Intra-lab repeatability	10. Intra-lab repeatability	10. Intra-lab repeatability
11. Comparison of existing reference methods	11. Comparison of existing reference methods	11. Comparison of existing reference methods
12. Collaborative study[a]	12. Laboratory reproducibility[b]	
13. Package insert review	13. Package insert review	
14. Quality policy certification	14. Quality policy certification	

[a] As defined in AOAC guidelines for conducting collaborative study
[b] As defined in AOAC-RI policy on laboratory verification

What do AOAC validated methods mean to the laboratory?
- increases method user confidence
- increases confidence in the user of analytical data
- provides information on methods performance characteristics allowing the method user to make judgements about data quality

Guidelines

It is essential that the interlaboratory study should only be carried out after thorough preparation by the AOAC Associate Referee, the volunteer who initiates the topic in AOAC. The necessary steps are well described in the AOAC guidelines for interlaboratory study, which describes how to optimize a new or available method, develop the within-laboratory attributes of the optimized method, prepare a description of the method that is not subject to bias, and how to invite participation in the study, how to write instructions and prepare report forms for participants, and how to facilitate familiarization of the technique by sending practice samples.

Method status and publication
- adopted as Official Method
- publication in Official Methods of Analysis
- interlaboratory study report published in Journal of AOAC, or other scientific journal
- after 2 years member ballot on final action

Design (samples, range, laboratories)

In order to prepare the design of the study, possible sources of variability must be identified and included. Only laboratories familiar with the technique used should be invited to participate. In view of possible outliers, it is recommended to invite ten or more laboratories to participate.

This interlaboratory study is a method performance study, not a study of the laboratory/analyst

Samples must be homogeneous and tested for homogeneity, and should be coded at random, including the two or more blind replicates. A blank or negative control, and, if available, reference materials should be provided. Spiked materials are recommended for recovery study, incurred materials for residues study.

Obligations of participants

The participants in the interlaboratory study must analyze as indicated and follow the method exactly, including the number of determinations as instructed, and not more! Individual values and blanks should be reported, and raw data and graphs should be supplied.The organizer should be called if any problems occur. A complete report should be submitted.

Frequent causes for outliers:
- calculation errors
- reporting errors
- incorrect standards
- contamination of reagents, equipment, test, samples

What is learned?
- The use of standardized terminology for method performance characteristics
- The scope, applicability and limitations of analytical methods
- To apply criteria for obtaining adequate qualitative and quantitive data
- To plan the conduct of a study
- To customize the design of a study
- To conduct multiple and single point calibrations
- To determine selectivity
- To assess methods bias and variability
- To conduct a ruggedness test.

Changes in Official Methods
- editorial
- method extension
- analyte addition
- substansive revision

Supporting data information depends on extent of modification
All modifications go through the same levels of review as required for adoption

From the above, it is demonstrated that organizing/participation in an interlaboratory method performance study is a useful exercise, which may be costly and time-consuming. Therefore it is important to realize that there is much potential benefit for:

The Analyst: by being recognized as an expert in the publication of study and method, by contributing to a successful validation of a method which will probably be used worldwide, contributing and being exposed by interaction with peers.

The Laboratory: the personnel time and costs spent provide a contribution to common regulatory and industry standards, support laboratory accreditation, afford likely entry into international organizations for standardizing methods of analysis, and strengthen confidence in the validity of the methods used in the participating laboratory.

The AOAC Peer-Verified Methods Program

For more than 100 years, AOAC had but one methods program. In response to the needs of analysts and laboratories for methods that were clearly candidates for use, as demonstrated by performance in more than one laboratory but not necessarily validated by a full interlaboratory study, AOAC has developed the Peer-Verified Methods (PVM) system.

Peer Verified Methods Objective:
To provide analytical methods that have not been otherwise evaluated with a level of validation acceptable for some laboratory users

This is intended to provide a class of tested methods which have not been subject of a full interlaboratory study. Through a significantly lower resources-demanding process, this program provides for the rapid movement of methods into and recognition by AOAC and a level of validation for many methods which might not be otherwise evaluated. A Peer-Verified Method can always undergo full interlaboratory study at a later stage, if so desired.

At least six samples at three concentrations per matrix (including incurred samples where applicable and possible), usually with at least one near the expected/normal specification level, are required.

The originating laboratory and at least one additional, independent laboratory must participate in the study. Replicate analyses are required to help establish within-laboratory repeatability (precision). PVM methods are announced in Inside Laboratory Management, AOAC's monthly magazine.

The 10 steps of a Peer-Verified Method:

1. Author develops in-house performance data.
2. Method write-up in format.
3. Design of protocol for second laboratory testing.
4. Select independent laboratory and arrange testing.
5. Independent laboratory conducts method performance testing.
6. Independent laboratory submits results to author.
7. Author analyzes all data and sends to AOAC.

8. All method information to Technical Referee.
9. Technical Referee selects at least two expert reviewers.
10. Based on reviews, Technical Referee grants acceptance as PVM method.

PVM status and publication
- accepted by AOAC as PVM method
- published as individual methods, on demand; by mail, fax, Internet
- report of study in the Journal of AOAC INTERNATIONAL
- duration of status; 5 years

Peer-Verified Methods may be modified as needed by users; modifications, with supporting data, may be submitted to AOAC and will be published as 'Notes' attached to the original method.

The AOAC Performance-Tested Test Kit Program

The AOAC Performance-Testing Program is designed essentially for test kits, to provide third-party verification of performance claims for commercial, proprietary test kit analytical methods. Operation of this program is under the AOAC Research Institute, a subsidiary of AOAC INTERNATIONAL.

Test Kit Performance Testing Objective
To provide third party verification of performance claims for commercial, proprietary test kit analytical methods

A set of requirements for verification of test kit performance has been developed and applied successfully. In 1996, because of financial constraints, the program was suspended temporarily and is now again in full operation.

Test kit producer-generated data to verify performance claims are required according to an AOAC-approved protocol, to generally include: ruggedness tests, calibration curves, accuracy, precision, cross reactivity, test kit component stability, detection limit, limit of quantification and rates of false positives and false negatives.

The verification of the producer's data on selected parameters must be performed by an independent laboratory, according to AOAC-approved protocol.

When the AOAC Research Institute grants the one year 'Performance-Tested' certificate, the AOAC RI Performance-Tested mark may be used under specific conditions.

Ten steps for the Test Kit Performance-Testing procedure are:
1. Kit producer prepares required package of information.
2. Producer submits application and application fee.
3. AOAC RI Staff performs preview.
4. Identification of Expert Reviewers and independent testing laboratory.
5. Expert Reviewers develop testing protocol.
6. Independent testing.
7. Expert Reviewers assess results.
8. Package insert review.
9. Certification.
10. License to use Performance-Tested mark for 1 year.

Changes in Performance Tested kits
- graduated levels of re-validation study
- same levels of review as required for certification

The test kit is certified as an AOAC Performance-Tested Kit and carries a licensed-to-use certification mark, which is granted for 1 year.

Notice of certification and validation information is given in Inside Laboratory Mangement, AOAC's monthly magazine.

At present, a special Task Force in AOAC INTERNATIONAL is studying the integration of the three methods validation systems. The reason for doing this is the increasing number of proprietary techniques that are submitted for validation. To ensure that this is done in a desirable way, with possibilities for yearly checking and updating of scientific advances and the validity of the technique itself and whether it is still on the market in the same format, a study is made on how best to integrate the systems. This may give opportunities for extensive validation by a full interlaboratory study in order to be compatible with other organizations.

The conclusions of the Task Force are discussed at AOAC meetings, and this may take some time. Information is available at AOAC INTERNATIONAL, Department of Technical Services.

Acknowledgement The author thanks Nancy Palmer, former Director of AOAC's Department of Technical Services, for valuable contributions.

References

1. J Assoc Off Anal Chem (1989) 72:694–704(1989)
2. The Referee (1996) 20:No. 10, 18–19

M. Lauwaars (✉)
AOAC INTERNATIONAL,
P.O. Box 153, 6720 AD Bennekom,
The Netherlands
Tel.: +31-318-418-725
Fax: +31-318-418-359
e-mail: lauwaars@worldonline.n

Chairpersons, members of relevant bodies, official authorities, scientific organisations and their aims; addresses and other useful information

Accred Qual Assur (1999) 4:533

Magnus Holmgren

Observing validation, uncertainty determination and traceability in developing Nordtest test methods

Nordtest has analysed and developed test methods for over 20 years and presently there are over 1250 Nordtest methods in 15 different technical areas. More than 520 of these methods have been developed in Nordtest projects while the rest are adopted and recommended for use by Nordtest. All test methods are listed in the *Nordtest Register of Test Methods* [1].

Requirements for test methods as well as the way they are developed are everlasting subjects of discussion. Examples of issues in connection with the development of methods are: validation of the method, measurement uncertainty and traceability of the method, documentation of the development work as well as endorsement, revision and withdrawal of a method. Literature dealing with these issues is published through many channels, e.g. the proceedings from EUROLAB Symposia [2–4]. There are several articles of interest in these proceedings for those who develop test methods.

There are, however, few reports which deal with all parts of the development process. In Nordtest technical report 403, "Observing validation, uncertainty determination and traceability in developing Nordtest test methods" [5], the issues mentioned above are analysed and suggestions of how to handle them are made. The analysis is not only limited to Nordtest methods but is generally valid for all types of test methods, e.g. in-house methods developed at one laboratory or international standards used world-wide.

The report is available from the Nordtest secretariat free of charge.

References

1. Nordtest (1998) Nordtest register of test methods. Nordtest Technical Reports and General Documents NT DOC, GEN 017. Nordtest, Espoo
2. (1992) Quality management and assurance in testing laboratories. 1st EUROLAB Symposium Proceedings. Strasbourg, France, January 28–30
3. (1994) Testing for the Years 2000. 2nd EUROLAB Symposium Proceedings. Florence, Italy, April 25–27
4. (1996) Testing and analysis for industrial competitiveness and sustainable development. 3rd EUROLAB Symposium Proceedings, Berlin, Germany, June 5–7
5. Holmgren M (1998) Observing validation, uncertainty determination and traceability in developing Nordtest test methods. NT TECHN REPORT 403. Nordtest, Espoo

M. Holmgren
SP Swedish National Testing and Research Institute, P.O. Box 857, SE-50115 Borås, Sweden
e-mail: magnus.holmgren@sp.se
Tel.: +46-33-16 50 07
Fax: +46-33-16 50 10